'Not without raising a wry smile, the
pheres of the planets to the air outside
Pr

'*Every Breath You Take* is one of that happy group of books that are written in an easily accessible style yet get across important facts about the world and what we are doing to it. Seemingly jocular in tone, Broomfield begins by dealing with atmospheres in general, on bodies in outer space, then closes in to deal with the structure of our own atmosphere. I learned many useful facts. But the book really comes into its own when he goes on to deal with the ways in which we are threatening the future of the planet, and of ourselves, by polluting the atmosphere. Amid entertaining slices of history about London smogs, the Clean Air Act and why the expensive districts of British towns are on the west side (to take advantage of a climate where prevailing winds are westerly), Broomfield hammers home a very clear statistic – air pollution causes seven million premature deaths per year, more than smoking, more than road accidents, while in the UK it is estimated to shorten the average lifespan by 6 months… Broomfield performs a very important service. Go out in your diesel car and buy this book, then read it while you huddle beside your wood burning stove.'

Peter Wadhams, author of *A Farewell to Ice* and Head of the Polar Ocean Physics Group, University of Cambridge

'Mark Broomfield's writing is just the breath of fresh air needed to lift the fog on atmospheric sciences and atmospheric chemistry in particular. His love of the research and the community he is part of shines through to ensure the book effortlessly links planet formation, air pollution, the ozone hole and climate change. Mark shows us that the air we breathe has no boundaries.'

Piers Forster, Professor of Physical Climate Change and Director of the Priestley International Centre for Climate at the University of Leeds

Mark Broomfield has a PhD in atmospheric chemistry, and has specialised in air quality, odour and health issues since 1992, both within the consultancy sector and as an industry specialist at ICI. He has completed research for DEFRA and Scottish Environment Protection Agency into the control of dioxins and furans, and carried out influential evaluations of the Government's Expert Panel on Air Quality Standards and Air Quality Expert Group. He lives in Shrewsbury.

EVERY BREATH YOU TAKE

A user's guide to the atmosphere

Dr Mark Broomfield

This edition first published in the United Kingdom by
Duckworth in 2019

Duckworth, an imprint of Prelude Books ltd
13 Carrington Road, Richmond TW10 5AA United Kingdom
www.preludebooks.co.uk
For bulk and special sales please contact
info@preludebooks.co.uk

A catalogue record for this book is available from the British Library

Printed and bound in Great Britain by Clays
ISBN 9780715653708

To Emma

CONTENTS

Introduction		1
Chapter 1:	Other Worlds, Other Atmospheres	11
Chapter 2:	Our World, Our Atmosphere	17
Chapter 3:	A Happy Atmosphere	48
Chapter 4	Ozone: Globally Helpful, Regionally Harmful	64
Chapter 5:	Whatever Happened To Acid Rain?	88
Chapter 6:	Air Pollution At The City Scale	100
Chapter 7:	Your Street, Your Neighbours, Your Home	155
Chapter 8:	Your Family, Your Body, Your Health	186
Chapter 9:	The Nuts And Bolts	217
Chapter 10:	The Future: My Life, My Car, My Town, My World	270
Notes		289
Acknowledgements		299
Index		301

INTRODUCTION

A JOURNEY IN SPACE

This is a book about a journey. Not a metaphorical journey, but not a real journey either: it's a theoretical journey through the atmosphere. And not just our friendly local atmosphere here on Earth, either: we're going to start outside the solar system, and take in some pretty bizarre and uncongenial atmospheres on our way to the outer reaches of our own planetary comfort blanket here on Earth. An atmosphere, as I hardly need to tell you, is *"the envelope of gases surrounding the Earth or another planet, the pervading tone or mood of a place, situation, or creative work."*[1] As a paid-up atmospheric scientist, I'm very much writing about literal atmospheres rather than pervading tones, so if you were hoping for a book about the mood of a place, situation or creative work, tuck this book under your arm, and head off to the Literary Criticism department to see if you can't pick up a "buy one get one free" bargain.

Welcome aboard! Our first stop will be Exoplanet GJ 1132 b – the "exo" means that it's outside the solar system, the GJ 1132 bit is the catchy name of the star it orbits around, and "b" means it's the first of GJ 1132 a's planets to be discovered (the star itself is assigned the letter "a"). As I write, there are four thousand confirmed planets outside the solar system,[2] and our friend GJ 1132 b is just a tiny bit more welcoming than the other 3,999 because it's the first Earth-like planet we've found with an atmosphere. Well, strictly speaking, it's the second. But before we get to the atmosphere that we live and breathe here on SUNb, we'll take the time to stop off at some of the other atmospheres in our solar system.

To be honest, the other atmospheres in our solar system are not very congenial: most of our neighbouring planets have thin, icy atmospheres, even colder than the atmosphere when you've just accidentally used your ex's name to refer to your partner. It was an accident! But we'll take a look at them nevertheless, and also drop in on Venus to see what global warming looks like when it properly takes hold.

And then we get to the fascinating, essential, fragile comfort blanket surrounding the Earth. From the outer reaches of the ionosphere, down through the thermosphere, the mesosphere and the stratosphere, we'll finally rock up at the place that we call home: the troposphere. Never heard of it? Don't worry, it's just the bottom few kilometres of the atmosphere, where we spend our lives, except when on international jet travel, so you've been breathing it for a long time (and even when flying, you're still breathing recycled troposphere). Within the troposphere, there's a journey to be had: from global pollution and climate issues through to regional and urban air pollution, and right down to the local effects of air pollution. Dust, smells, and the effect that air pollution has on the price of your house. Want to find out how to save 14% on the price of a house with a simple wave of a magic postcode? Read on.

TWO NUMBERS

Like Sellar and Yeatman's classic *1066 And All That*, this book contains two memorable numbers: if you forget everything else you read here, hold on to these. The first number, as I may have mentioned, is that air pollution is responsible for 14% of your house price. And the second number is that air pollution is responsible for seven million early deaths every year. I'm not sure which of those two numbers is the more memorable, but I do know which is the most shocking. Because now, after years of living in the shadow of the big kid on the block, climate change, air pollution is at last beginning to get the attention it deserves as we begin to grasp the magnitude of all those millions of unnecessary deaths.

What, you might ask, is an early death? That's a good question, and the short answer is that we don't exactly know. Spoiler alert, but everyone dies eventually, and so the result of environmental influences on health like air pollution is not to increase the rate of deaths per person (it's already 100%), but to shorten our life expectancy. We can describe this in a number of different ways. One way is to estimate the number of early deaths thought to occur each year – here in the UK, it's about forty thousand, and worldwide, it's about seven million. Another approach is to look at the overall effect on life expectancy: in the UK, air pollution reduces everyone's life expectancy by six months, on average. What we don't know is whether everyone experiences a comparable shortening of life, or whether the effects are even more dramatic in a smaller number of people. And we certainly don't know who all the people are who are dying early from too much air pollution every year. You might well suspect that the health of someone in your family has been affected by air pollution, and you might well be right, but no doctor has put "air pollution" as a cause of death on a death certificate. To be fair, that's mainly because death certificates record the conditions affecting individuals rather than speculating about possible external causes – but even if they did speculate, we can't identify individual deaths caused by poor air quality. That may be about to change, but for now, we just know that pollution, on average, makes people, on average, die a bit earlier.

And it's a really big deal. Globally, air pollution causes more early deaths every year than passive smoking, and obesity, and water pollution. Put together. Maybe obesity and passive smoking are more shocking, because they are both more tangible and more avoidable. The deep fried Mars Bar from the chippy on the corner, and a packet of fags from the paper shop next door are both real, physical objects, and we can choose to indulge in them, or to spare our arteries and our families from their long-term effects. In contrast, water and air pollution are not so easy to visualise. And there's a limit to what we as individuals can do to mitigate or avoid them. That's the thing about environmental problems: they are almost entirely caused by other people, and they quite often manifest themselves as a small risk or small effect on many people that you can't really spot at an individual level. But it turns out that air pollution is even more important than obesity, passive smoking and other A-list causes of early death. Every few seconds, you have to take a breath: you don't have much choice about what you breathe, and the air goes deep into your body where it's needed to keep you alive, but might also do you a little bit of harm.

A JOURNEY IN TIME

Such thoughts were a long way from my mind as I emerged blinking from the Cambridge University chemistry labs in the autumn of 1987 – indeed, it would be decades before anyone got to grips with the impact that air pollution has on the health of people everywhere. I was in the third year of my chemistry degree, and had decided that my rather limited talents did not lie in the field of organic chemistry or inorganic chemistry (these being, unfortunately, the areas of chemistry where money is most likely to be made). I was much more at home with physical and theoretical chemistry, and so I headed off to the first lecture in the first of four physical chemistry units available to final-year undergraduates on the obscure subject of the chemistry of atmospheres. Professor

Brian Thrush started the course by talking about atmospheric kinetics – that is, measuring how fast reactions proceed in the gas phase. An obscure corner of an obscure subject it might have been, but I remember walking out of that lecture on what was probably quite an ordinary November morning, probably raining, surrounded by students dressed in authentic 1980s retro, thinking "this is what I want to do." As Damascus Road conversions go, there are probably more dramatic stories (St Paul for one), but it was genuine and long-lasting: the atmosphere has stayed with me, and I've stuck with the atmosphere, over the three decades since then.

The course went on to look beyond laboratory measurements at the atmospheric reactions which govern photochemical ozone at ground level, and ozone in the stratosphere. Looking back, some of this was pretty cutting edge in 1987, just a couple of years after the discovery of the stratospheric ozone hole, for all that some of the lecturers might have given the impression that they had been giving the same lectures since 1947. The course went on to look at the atmospheres of other planets in a short series delivered by the always rigorous and entertaining David Husain. The final series of lectures addressed atmospheric monitoring techniques, and was delivered by a young-looking research fellow, Dr John Pyle – the same Professor John Pyle CBE FRS who went on to become head of the Department of Chemistry at Cambridge.

Well, that was it for me. Apart from a short diversion when I wondered if I ought to get involved in a social work of some kind (would have been disastrous, a lucky escape for all concerned), I headed off to do my PhD on measuring the spectra and atmospheric kinetics of reduced sulphur compounds. These chemicals are among the most smelly substances known to man, and the odour of dimethyl disulphide in particular is enough to make you want to curl up and die. But back in 1989, the old Central Electricity Generating Board which ran the UK's power stations wanted to know whether acid rain in Scandinavia could perhaps be caused by organic sulphur compounds released by micro-organisms in the North Sea. Never mind that these micro-organisms had presumably been chugging away releasing organic sulphur compounds since time immemorial – the CEGB wanted more information, and was prepared to pay a PhD student a bit of cash to do some basic research. The problem of acid rain in central and northern Europe was acute in the 1980s, and the CEGB was looking for possible explanations of the observed impacts on forests and lakes in Scandinavia and Central Europe. So they funded a research programme to develop a model of organic sulphur compounds in the atmosphere, with the aim of calculating the possible contribution of natural sources to acid rain. My part in this was to measure reaction rates for some of the chemical intermediates in the atmosphere – data which wasn't previously available.

To be honest, I'm not sure if the CEGB ever completed its modelling study. As I started my PhD, the CEGB was in the process of being privatised, and I got the impression that there was perhaps more attention being paid to transfer of terms and conditions and future pension arrangements (and probably the long-term reliability of the UK's electricity supply, to be fair), than to a speculative research programme and a lowly student in far-off York. But I completed my PhD, with the occasional visit from my industrial supervisor at CEGB, and with my academic supervisor Chris Anastasi and our Danish partners from the Risø National Laboratory (who helpfully provided apparatus which actually worked), published a bit of data to add to the sum of human knowledge. And as is the way with academic papers, it's still out there,[3] with a citation in the depths of NASA's magisterial *JPL Publication 15-10: Chemical Kinetics and Photochemical Data for Use in Atmospheric Studies,* ready for anyone who is interested in modelling organic sulphur reaction pathways in the atmosphere.

I was lucky. I first got interested in the chemistry and science of the atmosphere in the late 1980s, at an exciting time when the atmospheric science community, which is a thing, was investigating new problems and finding new solutions. Laboratories had exciting-looking lasers in them – well, some laboratories did; I had to make do with an ultraviolet lamp. And as part of that voyage of discovery, it turned out, no big surprise, that acid rain was caused by combustion of fossil fuels, at least partly in British power stations, so my study of organic sulphur compounds was, in the end, a particularly smelly red herring.

After the PhD came my first proper job (not counting waitering for a pound an hour plus tips – yes I know, that's daylight robbery, no way was I that good) as an environmental consultant specialising in air quality. Thirty years later, here I still am, not still waitering, fortunately for the customers who for all I know may still be waiting for a toasted teacake in Buddies tearoom in Chipping Ongar high street, but still dealing with what's in the atmosphere and what it does to us. One of the advantages of working on air pollution is that people are very often interested in the kind of work I do – at least for a minute or two, anyway, until they find out how much of it is sitting at a desk rather than measuring air quality in moorlands in Cornwall (though I have done that) or climbing up chimney stacks in winter snowstorms (done that too). At least "I'm an air quality consultant" usually gets a more positive response than "I'm a management consultant," even if it doesn't get a better pay scale.

Doesn't matter. I'd be as good a management consultant as I was a waiter. I love the dynamic activity in the atmosphere, driven by sun, wind and chemical reactions. I'm fascinated by the effects that air pollution has on health and natural ecosystems. It's science in the real world: a subject where you can use models to predict concentrations to eight decimal places, but which are sometimes wrong by a factor of ten. You can spend hundreds of thousands of pounds on measurement instruments, or a tenner on a small plastic tube. It's sometimes controversial, frequently frustrating, occasionally repetitive, and really, really important. Did I mention that seven million people die each year from air pollution? One day it's state-of-the-art instrumentation to measure air pollution in real time, the next day it's sniffing to see if we can detect a smell. It's lots of small decisions that all of us take which add up to a whole big load of air pollution – and the occasional big decisions which sometimes result in a lot of extra air pollution, but sometimes add up to no pollution at all.

THIS JOURNEY

Doesn't make sense? That's not surprising, because not many of us know a lot about the atmosphere. Come with me on my journey through atmospheres and through this layered, dirty, life-giving atmosphere of ours. We'll start the small matter of four hundred trillion km away at our new friend GJ 1132 b, and with each chapter get closer and closer to the Earth. There are some bizarre and strange atmospheres out there, from distant GJ 1132 b to the furnace of Venus. We'll find out what's in our own atmosphere: the good (life-giving oxygen), the bad (inert nitrogen and noble gases which would suffocate us if they could) and the ugly (the many different pollutants that we add to the mix, in surprisingly small quantities). At a global scale, we'll investigate the links with air pollution's big brother – climate change – and take a long, hard look at the enigma that is ozone. Then we'll take a look at some of the effects that air pollution has – on our health, on ecosystems, and on our senses. We'll look at some controversial questions, and how air pollution is reported and presented in the media: surprisingly, sometimes you can believe what you read in the papers. We end up (in Chapter 8) going right into our lungs to see what the atmosphere is doing to you and me right now. Then we turn around to take a look at some of the dark arts of dealing with air pollution – how all those nasty pollutants behave in the atmosphere, and how we study them, before looking at what the future holds for the atmosphere, by which I include the air and the pollutants that we seem addicted to putting into it: all the stuff that ends up in every breath you take.

This is an exciting time to find out more about air quality, and look forward perhaps to building on the success stories of the past half-century to improve air quality in the many places where levels continue to be way too high, and in some places still going up. I've written this book with the aim of passing on some of my enthusiasm for what I think is a fascinating and very important subject. Dealing with air quality brings together science, politics, economics and psychology, and probably a few other -ologies as well. The precise world of scientific measurement and evaluation meets the messy world in which we all have to live. And the effects of air pollution on the health of millions of individuals around the world are real and important. I've included references at the end of the book for anyone who's interested in checking up or following up.

"Every breath you take, I'll be watching you," wrote Sting for The Police in 1983. At the time, I thought it was a love song, but it turns out that he's talking about the behaviour of an obsessive stalker, maybe following the break-up of a relationship. As Sting himself said, *"It's about jealousy and surveillance and ownership."*[4] There isn't room for ownership of the actual breaths that we take – at least, not until someone tries to privatise the atmosphere. But there is definitely a place for surveillance of what's in the air – and, as we'll see, maybe even a place for jealousy of the good air quality some of us get to enjoy with every breath we take.

CHAPTER 1

OTHER WORLDS, OTHER ATMOSPHERES

Within Four Hundred Trillion Kilometres

A PLACE LIKE HOME

So we start our journey thirty-nine light years away. That's a mere four hundred trillion kilometres, a distance that you could theoretically cover in – well, thirty-nine years, if you went as fast as the speed of light. The record for the fastest speed attained by a manned vehicle is held by Apollo 10, at 11.1 kilometres per second – decidedly frisky, but some way off the speed of light. At Apollo 10 speed, it would take you a shade over a million years to get there. No wonder the Starship Enterprise used warp drive to get around the universe.

It's only in the past few years that we have started to make serious progress in discovering planets outside the solar system. Three quarters of the planets that we know about have been discovered since 2013 – indeed, 40% were discovered in 2016 alone. So we haven't had long to find out about any atmospheres. One of the exoplanets discovered in 2015 was a small, rocky object called Gliese 1132 b, known for short by the snappy nickname of GJ 1132 b. These astronomers and their crazy informality. This tiny dot in space is a rocky planet, a bit larger than the Earth, hurtling round a red dwarf star about four times a week. In April 2017, a team from Cambridge

University and the Max Planck Institute for Astronomy reported that GJ 1132 b has an atmosphere: this was the first identification of an atmosphere of any kind, surrounding an Earth-like planet, apart from the one that you're breathing right now. We do have to be a bit cautious here. No one's popped off for a couple of million years to take a sample of the atmosphere: instead, the atmosphere was identified by looking at how the planet dims the light of its star as it passes between the star and the European Southern Observatory in Chile. *"An atmosphere rich in water and methane would explain the observations very well,"* the team members said.[5]

So it looks likely that GJ has an atmosphere, and that atmosphere has the kind of stuff in it that makes the researchers think it's going to be hot there. Really hot. To be specific, a surface temperature of 370 degrees centigrade, which would make it uninhabitable for any life form that we know on Earth. An atmosphere made up of greenhouse gases with a red-hot surface temperature inevitably reminds me of Venus... which we'll come to shortly.

But first, looking at the atmosphere of a planet outside the solar system made me wonder about a question that I'd never really thought about. Where do all these atmospheres come from? The answer seems to be mainly from within the planets themselves. A small, rocky planet like Earth or GJ 1132 b starts out its life as an accumulation of solar dust close to a star. The heat from the star vaporises most substances, leaving hot minerals behind. These materials slowly stick together and, as they cool down, there is a ferocious amount of geothermal activity. All sorts of gases pour out of volcanoes and into the atmosphere – carbon dioxide, hydrogen sulphide and water vapour are very popular components in your nascent atmosphere.

INTO THE SOLAR SYSTEM

Things are different on gaseous planets like Jupiter and Saturn. These planets were formed from material in the more distant, cooler parts of the solar nebula, which means they're mainly hydrogen and helium. They're pretty much gas all the way down – or putting it another way, these planets are 100% atmosphere. So let's stop off at Saturn on our way in towards the Earth and specifically your house. Saturn is less than a half-millionth of the distance to GJ 1132 b away, so we're already over 99.9995% of the way there.

Because it's made of left-over bits of sun, Saturn's atmosphere, and indeed Saturn itself, is 93% hydrogen and 7% helium. Further down, there's a lot more helium in the atmosphere, suggesting that the heavier helium is sinking towards the centre of Saturn. There isn't really any distinction between the planet and its atmosphere, so to enable us to talk about the gas giant planets in the same way that we consider the atmosphere of rocky planets, we define the surface of Saturn and its neighbour Jupiter as the point at which the "air" pressure is the same as it is at the Earth's surface. Helpfully, that's known as "1 atmosphere". Saturn's atmosphere contains small amounts of ice crystals and sulphur, and some howling winds at over a thousand miles per hour. All in all, if you're standing on Saturn at the equator, it's blowing a hydrogen gale, there's a smell of sulphur, it's almost 200 degrees below zero, and there's nothing to stand on.

So let's pop down the road to Jupiter, conveniently located 43 light minutes from the Earth at its closest. Not quite as cold, not quite as windy, but just as hydrogen-y and helium-y. Jupiter famously features a red spot to the south of the equator, which is a hellish, spinning storm. It's been going for at least 350 years (we think – it depends on how you interpret observations from the early days of telescopes), although it seems to be shrinking, so maybe the storm will come to an end at some point in the next few hundred years. Just like Saturn, Jupiter's atmosphere has trace amounts of ammonia, hydrogen sulphide, water and methane, which result in the bands of colour we can see running parallel to the equator. Some of these chemicals are pretty smelly, but of course you'd be too busy gasping for breath to worry about the smell of ammonia or hydrogen sulphide. Both Jupiter and Saturn have a temperature structure within their atmospheres which, to me, is surprisingly similar to the temperature profile of the Earth's atmosphere. The temperature of Jupiter's atmosphere drops as you go up from "ground level" to a height of about fifty kilometres. After that, the temperature changes direction, and starts to increase as you get higher, up to a height of 200 kilometres, where it reaches a sultry minus 100 degrees centigrade. At that point, things change round again, and start getting cooler as you get higher, although there's less and less of anything to actually be cold. We'll look at the reasons for this in the next chapter: when we get to the Earth, we'll find that there are intimate connections between these temperature changes, and some of the biggest air quality stories around.

THE NEIGHBOURS

Before we get to the Earth, we're going to stop off at Mars, which can be a mere thirteen light minutes away. At our pedestrian top speed of eleven kilometres per second, it's going to take less than a year to get there from the Earth. Now, that's beginning to sound more possible, so what kind of atmosphere can we expect to find when we get there? The answer is: almost, but not quite, nothing. Mars has an atmosphere which is about one hundredth the density of the Earth's. But at least it has a rocky surface that we could stand on if we wanted to. The thin air of Mars consists mainly of carbon dioxide, with a couple of percent each of argon and nitrogen. It seems that Mars used to have a lot more carbon dioxide in its atmosphere, presumably coming from geothermal activity as the planet cooled, but this has been lost – maybe by the influence of the sun, maybe following a catastrophic impact from a smaller body. While Mars's atmosphere doesn't seem to have too much to offer us, it does have two atmospheric phenomena which are pretty familiar: dust storms and snow. The familiar red dust of Mars (red because it's made of iron oxide) is kicked up into huge dust storms on a global scale, which can last for months. And what about that snow? Well, sorry to disappoint, but while snow or fog does sometimes feature in the Martian weather forecast, it's not the watery stuff that we're familiar with, but flakes of solid carbon dioxide which fall, particularly at the poles. It looks like Mars used to have a lot of water on and below the planet's surface, so it's possible that there was once, to coin a phrase, life on Mars. But those days have gone, along with the atmosphere that, perhaps, used to protect the red planet's water resources. Nowadays, while there is plenty of water left on Mars, it's in the form of ice at the planet's north and south poles. Could we, or any other slightly familiar form of life, make use of that?

Well, let's leave that question hanging, as it were, in the air, and move on from the sterile atmosphere of Mars, past the Earth for a minute, and take a look at Mercury and Venus. Mercury's the easy one: it's so close to the sun that its atmosphere has more or less been stripped back to almost nothing. It's the thinnest atmosphere in the solar system, with a pressure of less than one trillionth of the Earth's atmosphere at the surface. What little stuff there is comes from the solar wind, or bits and pieces released from collisions of micrometeorites with the surface. Doesn't really matter, because any particles which form are removed by the solar wind and by the magnetic field of Mercury itself. The temperature on Mercury varies wildly because it's so close to the sun, and there isn't an atmosphere to smooth out variations in heat striking the planet. Really, really cold if you're facing away from the sun, and unbelievably hot if, by some bizarre chain of events, you should find yourself on the side of Mercury facing the sun.

Hopping one out from Mercury brings us to Venus, and what a mysterious place it is. It would be a terrible place to visit: the hottest place in the solar system, with an atmosphere made up almost entirely of carbon dioxide, plus a few percent of nitrogen. Carbon dioxide is a greenhouse gas, which means that the atmosphere of Venus soaks up the sun's energy without letting it go

– and that's what makes it so very, very hot. Clouds of sulphuric acid provide a bit of acrid variety to the hot, choking mix of inert gases. And what an atmosphere it is: with a pressure at the surface almost a hundred times the pressure you'd experience on the sea bed at a depth of three thousand feet. Except hotter. We can't yet be sure why Venus's atmosphere has developed in the way it has, but maybe a large quantity of water evaporated from the surface in the early years of the planet's life.[6] Water vapour is also a greenhouse gas, so this damp atmosphere would have started getting hotter and hotter, and this in turn may have prevented carbon dioxide forming geological rocks as carbonates. All this carbon dioxide ended up in the atmosphere, so life on Venus just carried on getting hotter and hotter – not that there is any life on Venus, so far as we know.

And finally, just round the corner from Venus, we reach our highly amenable planet Earth. With its atmosphere that's not too hot, not too cold, got a bit of oxygen but not too much, Earth's atmosphere appears to be pretty much ideal for life as we know it. Now, of course, that's a self-fulfilling prophecy: any planet on which we have evolved would have a lovely atmosphere that would suit us just nicely. We certainly haven't yet found anything that would even come close as a home for life. This doesn't mean there isn't one out there somewhere, but it does mean that our blue-green home with its life-giving atmosphere is really, really special, and definitely worth looking after.

CHAPTER 2

OUR WORLD, OUR ATMOSPHERE

Within Forty Thousand Kilometres

MOST OF OUR ATMOSPHERE

We've been on quite a journey to get here, but at last we've ended up on Earth. You're never more than 20,000 kilometres from home on Earth, which is perhaps a comforting thought. We share our perfectly proportioned life-giving atmosphere with, among other life-forms, eight billion people. And all of us are breathing. In, out. In, out. We literally can't get enough of it. So what's in this atmosphere of ours that makes it so great?

Perhaps surprisingly, "every breath you take" here on Earth contains about four-fifths nitrogen. Now, nitrogen is not a gas with a lot of pizzazz. We breathe it in, and then we breathe it out again. Nitrogen is an important part of our lives in many ways – helping plants to grow, for example – but it doesn't get up to much in the atmosphere. There's also a sniff of argon, which is even more undistinguished than nitrogen. As we'll see, argon belongs to the group of chemicals known as "inert gases" which says all you need to know about what happens to argon when you breathe in. Nothing. You breathe it straight out again.

With so much nothing going on, it took a while for scientists to get to grips with what's in the atmosphere, or indeed, to understand that the air was more than just nothing, what with it being invisible as well. To be fair, chemistry itself didn't start to shake off the shackles of alchemy, its disreputable and embarrassing uncle, until the seventeenth century. Yet by 1774, only about a hundred years after the first wobbly steps had been taken in this new scientific way of analysing the world around us, natural philosophers had identified 99% of the atmosphere – both nitrogen and oxygen. I think that's pretty good going, especially when the modern concept of a gas hadn't even been invented until the publication of Jan Baptist van Helmont's *Ortus medicinae* in 1648. It was important for our emerging understanding of the atmosphere to get these basics sorted out, because all the action takes place in the one-fifth of the atmosphere which isn't argon or nitrogen. We now know that this is almost entirely made up of oxygen. Almost entirely. In fact, pretty much the whole of this book is contained in that little word "almost". What I'm interested in is the very small proportion of our atmosphere that isn't nitrogen, argon or oxygen. But before we get on to that, let's take a quick look at oxygen, which at least does more than try to asphyxiate us.

Oxygen came close to being named "fire air", "vital air" and the unbeatably obscure "dephlogisticated air", before ending up being called oxygen, which means, rather inaccurately, "acid-forming". I do wish it was still called "vital air", but at least we've got away without having to call it "dephlogisticated air". There's plenty more about phlogiston in Chapter 9, where we'll find out why oxygen nearly ended up with this unwieldy name. "Vital air" (or maybe "vitalium") would have been spot on, because oxygen is what brings us life. And not only does oxygen bring us life, but it is life that brings us oxygen. When the Earth was formed, four and a half billion years ago, there wasn't any oxygen in the atmosphere. Indeed, we don't yet know of any other planet with oxygen in the atmosphere. It wasn't until single-celled organisms started producing oxygen after half a billion years or so that there was any chance at all of oxygen in the atmosphere. But in fact, the atmosphere remained oxygen-free for another billion years, as all the oxygen produced by the earliest plant life was gobbled up by hydrogen, or by reacting with iron and other geological materials.[7] This went on for about a billion years before photosynthesis started to get under way, resulting in enough oxygen production to enable it to begin building up in the atmosphere at the same time as the production of hydrogen from volcanoes started to drop off.

From about 2.3 billion years ago onwards, oxygen levels in the atmosphere started to climb up to about 3%, and stayed there for a very long time. A very, very long time actually: right up to 700 million years ago or thereabouts. After that, oxygen levels started going a bit crazy. A steep climb up to 13% in the blink of an eye – a mere fifty million years or so – then another sudden increase to the highest ever level[8] in the Carboniferous period about 300 million years ago. By then, oxygen levels were 50% higher than they are now. This high level of oxygen resulted in the development of some remarkable life forms - dragonflies the size of seagulls, spiders with legs almost half a metre long, and centipedes extending a metre from one end to the other.[9] These frankly terrifying bugs were only possible because of the higher levels of respiration available as a result of an atmosphere containing over 30% oxygen.

The reason for this increase in oxygen levels around 300 million years ago seems to be the very rapid burying of forests, which contained huge quantities of carbon. Removal of carbon meant that oxygen produced by photosynthesis had nowhere to go except into the atmosphere, where it was able to build up to about 35% at its highest. Still, good old mankind: never happy to let well enough alone, we've been doing our worst to redress this balance for the last couple of hundred years, by digging up these old forests, conveniently turned into coal and oil for our benefit, and burning them to produce heat and electricity.

As far as we know, the Earth is unique in producing oxygen. So while there might be other planets out there which fulfil some of the criteria for life to evolve, we certainly haven't found any with evidence of oxygen-based life. Indeed, if we do ever find another oxygen-rich atmosphere, this might be a strong indication that life has evolved on that planet and is busily photosynthesising.

Back on Earth, while plants are doing their best to put oxygen into the atmosphere, we animals are doing our best to suck it out again. As we breathe in, some of the oxygen in the air is transferred into the bloodstream through the enormous surface area of our lungs. The total surface area of adult human lungs is about fifty to seventy square metres, or roughly one side of a tennis court, but please don't try to verify this estimate at your local tennis club. This large surface area enables our bodies to extract about a quarter of the oxygen present in every breath you take, so when we breathe out, the air contains about 15% oxygen. In, 21%; Out, 15%. Something similar happens when fuels are burnt. Like respiration, the process of burning a fuel is essentially converting carbon and hydrogen in the fuel by the addition of oxygen to result in carbon dioxide and water. The principle is the same whether it's an internal combustion engine in a car or lorry, a log burner in the sitting room, the latest gas-fired power station, or your lungs and bloodstream. And there's a lot of that kind of thing going on. In fact, what with all that breathing, and burning fossil fuels, we are extracting oxygen from the atmosphere a bit faster than plant life can put it back in again. This means that oxygen levels in the atmosphere are currently decreasing by about 0.0019% per year.[10] At this rate, in 500 years, oxygen levels in the atmosphere will have dropped from 21% to 20%. Not enough to worry about, yet, but maybe worth keeping an eye on.

GASPING FOR BREATH

If you're interested in experiencing life in a lower-oxygen environment in a quicker way than waiting for a couple of millenia, an alternative approach is to climb up into the mountains. While writing this book, I travelled to Nepal and had the privilege of crossing the Larke Pass to the north of Mount Manaslu in the Nepal Himalaya. The Larke Pass goes up to about 5,100 metres above sea level, which is plenty high enough to experience changes in oxygen levels.

At this height, air pressure is about half what it is at ground level, so not surprisingly, I found myself breathing hard when engaged in strenuous activities such as doing up a shoelace or standing still. Although I struggled a bit, it's still amazing that someone like me, who's spent the past fifty years living within a couple of hundred metres of sea level, could adapt to the altitude sufficiently to walk across the pass, and even smile for the camera, rather than just gasp feebly for breath. Yes, that is a smile. This experience brought home to me how amazingly adaptable our bodies are, and equally how fragile and small the atmosphere

The author, 5106 metres above sea level at the Larke Pass, Nepal[11]

is. Over a couple of weeks, I was able to walk from the familiar comfort of one atmosphere pressure to the, occasionally severe, discomfort of half an atmosphere, and then in a couple of days, return to more or less normal conditions. Normal for me, anyway. Our Nepalese team were able to walk cheerfully up and over the 5,100 metre pass, carrying 25 kilogram loads on their backs, and still come scampering back with flasks of tea for us more ponderous Westerners. It's all a matter of what you're used to, and we can get used even to half as much oxygen as we get at sea level.

Throughout the walk around Mount Manaslu, we encountered prayer flags at particularly significant points, like those shown in the photograph. The flags are coloured blue, white, red, green and yellow, definitely in that order, representing the sky (and space), wind, fire, water and earth. The Buddhist culture in this part of Nepal traditionally distinguishes between sky and wind. Although I'd never thought about this before, it seems entirely reasonable to me, as the sky is clearly blue and out of reach, whereas wind is invisible yet at the same time, often very tangible. When you're halfway up the side of the eighth highest mountain in the world, I can confirm that the wind is very tangible indeed. However, both the blue sky and the invisible wind are made up of the same stuff, so why is it that the sky appears blue to us, whereas the atmosphere is transparent? The answer is that the blue colour of the sky derives from scattered sunlight. As long as we're not looking directly at the sun, which I'm sure we aren't for safety reasons, our perception of the colour of the sky derives from sunlight which is scattered as it passes through the atmosphere. The extent of light scattering is inversely proportional to the fourth power of the wavelength of the light, so shorter wavelength light is scattered much, much more than longer wavelength light. As blue light has a shorter wavelength than green, yellow, orange or red light, it is scattered more strongly – up to ten times more strongly when compared to red light. So most of the light that we see when we look up at a clear sky is blue, because of the high intensity of shorter wavelength colours – blues and purples. As a result, we perceive the sky as having a blue colour. At an altitude of 5,100 metres, there was half as much atmosphere above my head, so the intensity of light scattering of all wavelengths was lower than I'm used to down at sea level. This made the sky darker than usual for a bright spring day, and an even deeper blue, heading towards a dark blue or inky purple as I looked further away from the sun.

LAYERS IN THE ATMOSPHERE

The atmosphere is estimated to weigh about 5.5 quadrillion tonnes, or one millionth of the mass of the Earth.[12] Half of everything in the atmosphere was below my feet when I climbed to five thousand metres above sea level. The atmosphere is impossibly small, spread like a single layer of varnish over the Earth's surface. It's not only small and fragile, but on top of that, we kind of need our planet's atmosphere to stick around, so we really, really should be looking after it.

Even at the top of the Larke Pass, I was only about halfway up through the lowest layer of the atmosphere. We usually divide the atmosphere up into four layers – from the ground up, these are the troposphere (where weather happens, as well as being where people live, birds fly and middle-aged men succumb to their midlife crises by trekking in the Himalaya), the stratosphere (which includes the ozone layer and long-distance flights), the mesosphere (shooting stars) and the ionosphere (the Northern and Southern Lights). The dividing lines between these layers are not just arbitrary divisions: they have a real, physical basis in sudden changes in the temperature profile of the atmosphere.

Starting at the bottom, the troposphere contains three quarters of the atmosphere, and it's where almost all the weather happens – things are a lot quieter higher up. The troposphere is highest at the equator – about sixteen kilometres above ground level here – and gets lower towards the poles, where it extends only up to about 8

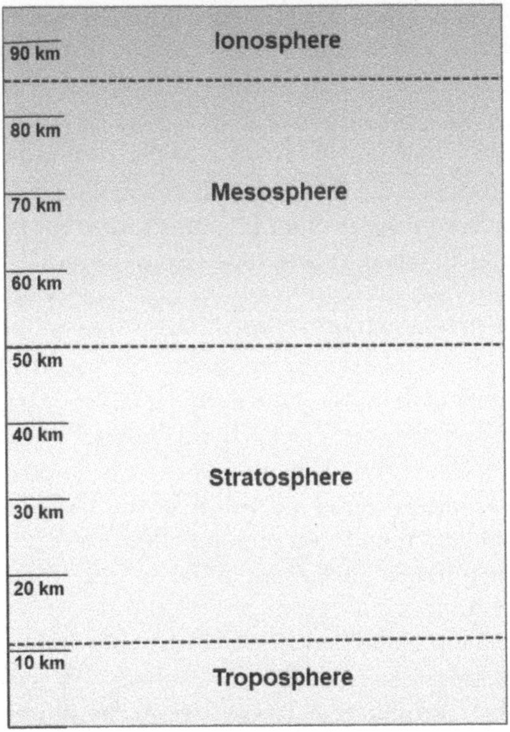

The four layers of the Earth's atmosphere

kilometres above ground level. As you go upwards from ground level, the temperature drops by about 6.5 degrees centigrade per kilometre. This temperature drop is caused by the expansion of gases as they rise through the troposphere.

At the top of the troposphere, there is an abrupt change to this profile. The cooling down of the atmosphere suddenly gets slower, and then reverses so that the temperature starts to increase with height. The top of the troposphere (known as the tropopause) is usually defined as the point at which the rate of change of temperature with height is less than 2 degrees centigrade per kilometre and the temperature here is typically around -60 degrees centigrade. This change in temperature profile effectively puts a lid on the troposphere. Where colder gases are located above warmer gases, mixing can take place as the cooler air sinks and the warmer air rises – or, if you insist on a bit more precision, mixing can take place where gases at a temperature higher than that which would result from vertical movement at the adiabatic lapse rate are located above cooler gases. Partly because of the troposphere's temperature profile, it is a fantastically dynamic place with mixing and movement of air masses, clouds, rain, winds, sunsets and everything else that familiarity has led us to think of as nothing more than a bit of weather. Of course, that bit of weather remains a big deal for us – not content with having its own show on the TV several times a day, the weather continues to be the go-to topic of conversation for most Brits. And why not? I know our slightly damp island doesn't experience much in the way of extreme weather, but on one day in Edinburgh recently, I experienced warm sunshine, strong winds, scudding clouds, torrential rain, double rainbows, and to finish it all off, a gloriously golden sunset. You can probably remember something similar. Got to love the troposphere. In contrast, where warmer gases are above colder gases, as is the case in the stratosphere, there is no thermally driven mixing. The lighter, warmer parts of the atmosphere, and the denser cooler bits, stay where they are. Unlikely as it seems, the stratospheric weather forecast is even more uneventful than the one we get on BBC *Midlands Today*.

What causes this abrupt change in temperature gradient at the top of the troposphere? It's a little chemical called ozone which turns out to be important to our understanding of how the atmosphere works, and indeed to our very survival. All that is yet to come, but for now, what matters is that ozone is a good absorber of sunlight. As ozone absorbs sunlight, it radiates longer wavelength infra-red radiation, also known as "heat". So where this process is going on, there is a steady absorption of sunlight and release of heat. At the top of the troposphere, ozone is formed by the action of ultraviolet light on oxygen. Ozone in this zone absorbs a bit more sunlight and emits a bit of heat in a cyclic process, before ultimately being removed by chemical processes. This results in a Goldilocks region for ozone (I wish we'd called it the "ozone zone" not the "ozone layer") where the balance between these processes are just right for a decent amount of ozone to be present. The cyclic process in which ozone absorbs ultraviolet sunlight and emits heat is responsible for putting the brakes on the decrease in temperature with height at the top of the troposphere and into the stratosphere. Because of this sudden increase in temperature, there is relatively little transfer between the troposphere and the next layer up: the stratosphere. This lid on the troposphere is not completely watertight and there are leaks between the troposphere and stratosphere (otherwise we wouldn't be worrying about making holes in the stratospheric ozone layer with those pesky chlorofluorocarbons), but it's a very limited process.

So crossing the tropopause brings us into the stratosphere. The stratosphere contains 90% of the ozone in the atmosphere: there is much more on the ozone layer in Chapter 4. Many commercial flights take place just within the stratosphere, where there is little turbulence and hardly any cloud, making for a smooth flight and good visibility down onto whatever clouds are doing their stuff in the troposphere below. Next time you're on an international flight, you can entertain your lucky neighbours by pointing out of the window and saying "See that? That's the tropopause, that is." Never fails.

As we continue to go higher within the atmosphere, the temperature stays pretty constant, but then starts to increase with height. By the time we get to the top of the stratosphere at about fifty kilometres above ground level, the temperature is back up to a steamy -15 centigrade. But it wouldn't feel very congenial out there, because the air pressure at the top of the stratosphere is only one thousandth of an atmosphere. At this point, the ozone and its warming effect runs out, and so we're back to the temperature of the atmosphere reducing as we move upwards into the mesosphere.

The mesosphere does a good job of protecting us from meteors and meteorites, which burn up on entering this part of the Earth's atmosphere. Even the extremely low air pressures in the mesosphere are enough to heat up and, generally, destroy objects falling towards the Earth's surface. We sometimes call these burning objects in the mesosphere "shooting stars", but as I hardly need to mention, they aren't actually stars – just vaporising meteors. The mesosphere extends up to about eighty or ninety kilometres above the Earth's surface. As I don't come across the mesosphere much during the course of my work on air quality in the lowest parts of the atmosphere, I thought I should try to find out a bit more about it. It turns out that the mesosphere is quite a mysterious part of the atmosphere, being too high to reach with aircraft or balloons, and too low to be investigated by satellites. Consequently, we know less about the mesosphere than about the atmosphere either side of it. The meteors which burn up in the mesosphere result in high levels of iron atoms up here. Under very cold conditions above the North and South Poles, ice crystals form around the dust left from burnt-up meteors, resulting in ghostly silver-blue clouds in the mesosphere. And it seems that scientists working on mesospheric geophysics take a quirky view of their esoteric subject matter, happy to discuss sprites and elves without any sense of irony – for evidence, I give you the 2003 paper from the University of Alaska: "Imaging of elves, halos and sprite initiation at 1 ms time resolution."[13] Sprites, elves and other manifestations are transient luminous events linked to lightning storms lower down in the atmosphere. Sprites go up and down (one variety is called a "carrot sprite"), while elves radiate outwards in the mesosphere. Elves is a not very accurate acronym for "Emission of Light and Very Low

Frequency Perturbations Due to Electromagnetic Pulse Sources" (ELVLFPEPS would have been closer), which makes me wonder if atmospheric physicists are perhaps a little more whimsical in their nomenclature than you might think.

Above the mesosphere, things get very strange. We're now into the ionosphere, and there's not much of anything up here. Temperatures increase again, due to the absorption of ultraviolet radiation from the sun, which is so packed with energy it can strip out an electron from an otherwise stable molecule, forming an electrically charged ion. The energy released as heat during these processes takes over from cooling due to falling pressure, and depending on the activity of the sun, the temperature of the mesosphere can get up to 2,000 degrees centigrade. Between about 90 and 130 kilometres above ground level, there is a layer of sodium atoms, formed from the action of atmospheric and frictional heat on incoming meteors and dust (about 30 tonnes of dust enters the Earth's atmosphere every day). High up in the ionosphere is where the Northern Lights and Southern Lights form, due to interactions between charged particles periodically released from the sun (so-called solar wind) and the atoms in the atmosphere. It's strange, and stranger still to think that all this strangeness is taking place quite close to where we live. If someone would only invent a vertical road, you could drive from your house to the top of the mesosphere in about an hour. By that time, 99.998% of the stuff in the atmosphere would be below you. And watch out if you keep going for another two or three hours further up into the ionosphere to 400 kilometres above the Earth's surface: you might bang your head on the International Space Station.

THE REST OF OUR ATMOSPHERE

Let's come back down to ground level, where the atmosphere is mostly more or less inert nitrogen, with about one-fifth made up of the more active and useful and life-giving oxygen. But what makes the atmosphere much, much more interesting to me is that the interesting one-fifth doesn't just contain oxygen. It contains much smaller amounts of all kinds of chemicals – the ones that smell, the ones that might make you feel wheezy, the ones that might turn a nice view into a hazy murk, the ones that react with each other... and the chemicals which mean that air pollution is responsible for seven million early deaths each year.[14] Yes, seven million. The World Health Organization looked at the effect of environmental factors on health worldwide,[15] and found that nearly one in four deaths are due to people living or working in an unhealthy environment. Environmental risks include air, soil and water pollution, climate change and exposure to chemicals and sunlight, and air pollution accounts for over half of deaths due to environmental factors. This makes air pollution about as deadly as tobacco (also responsible for about seven million deaths per year).[16] And it's more serious than not just passive smoking (just under one million deaths per year), obesity (three million deaths per year),[17] contaminated water (about half a million deaths per year)[18] or road accidents (1.3 million deaths per year),[19] but more serious than all of them put together.

All that from a tiny bit of one-fifth of the atmosphere! I think that's amazing, and worth knowing a bit more about. But what are these substances which have such a disproportionate effect? Well, before we start on the active bits, there are a few more inert substances to check off.

LAZY ARGON

Let's meet Professor William Ramsay of Edinburgh University, physical chemist and something of an expert discoverer of chemical elements, with a total of four to his name (that's pretty good going for one individual, out of a total of just one hundred and eighteen elements). Towards the end of the nineteenth century, Ramsay got on the track of another component of the atmosphere. Working with Lord Rayleigh, he found that nitrogen distilled from the atmosphere was slightly heavier than nitrogen obtained from chemical reactions. It was only a matter of half a percent or so, but that was enough to suggest that there was something else mixed in with atmospheric nitrogen. That something was an inert gas which he named "argon", which translates as "lazy". It turned out that lazy argon makes up almost 1% of the atmosphere. Ramsay and his partners went on to discover all the inert gases – helium, neon, argon, krypton, xenon and radon, which was such a big deal that it required a redesign of the periodic table to accommodate these new, and very lazy, elements. Want to know why inert gases are so inert? It's because chemical systems move towards their lowest energy configurations. The lowest energy configuration for individual atoms is one where the atom has complete layers of negatively charged electrons around the positively charged nucleus. Atoms can reach this state by nicking electrons from other atoms, or donating their electrons, resulting in a net negative or positive charged particle, called an "ion". Or they can also reach this state by co-operatively sharing electrons with other atoms, resulting in a chemical bond known as a "covalent" bond. Well, the inert gases aren't having any of that. They already have complete layers of electrons around the nucleus, so they wish to be neither borrowers nor lenders, nor indeed sharers, of electrons, thank you very much. They just sit smugly on the sidelines with their complete layers, watching all the other elements fight it out. Very much the Switzerland of atmospheric chemistry.

USEFUL HELIUM

There's another of the inert gases in the atmosphere: much smaller quantities than argon, but much more useful, and that's helium. Helium is the second smallest atom there is (after hydrogen), and its small size and inert nature are what makes it at the same time very useful and very rare. Helium is not only used in birthday balloons and giving you a squeaky voice if you breathe it in. It's actually a sensationally useful substance for running powerful magnets. And anyone who's ever had a Magnetic Resonance Imaging scan has cause to be grateful for the helium that makes the magnets work to produce the almost surreally detailed MRI images, without all that tedious cutting open and sewing up. Liquid helium is used to cool the magnetic coils down to a few tens of degrees above absolute zero. That's really, really cold: around -250 degrees centigrade. At such low temperatures, the wire becomes superconducting, and so very high currents can be passed through the coils. Such high currents result in the strong magnetic fields which are needed to produce sensational, and medically very useful, three dimensional MRI images.

But here's the thing. Helium is, perhaps surprisingly for such a volatile substance, a mineral product. It's produced in the ground by the radioactive decay of minerals containing uranium and thorium. Helium percolates up through the ground and is released into the atmosphere. Despite this reasonably significant source of helium, it only makes up about five millionths of the atmosphere. It would be a lot more, but helium is so light that a small fraction of the helium present in the atmosphere at any time can escape the Earth's gravity and disappear out into space somewhere. This is a slow process, but it's fast enough to remove helium permanently in the long term. So, although helium is the second most abundant element in the universe (after hydrogen), there's hardly any of it in the atmosphere. Fortunately, there are reserves of helium which have formed over millions of years, and been trapped under impermeable rocks – similarly to the way that natural gas reserves are formed. That means that we can obtain helium as a by-product from production of natural gas in areas where uranium and thorium are present. In some areas, such as the southern United States and (more recently) Tanzania, natural gas contains quite a lot of helium, which can be removed from the gas by distillation. Many more gas reserves contain a bit of helium, but at such low levels that it's not worthwhile for the gas company to extract the helium – so it is eventually just released to the atmosphere, to be lost into space. The new find in Tanzania is good news, and seems likely to provide a couple of decades' supply of helium, but helium is essentially a non-renewable mineral resource, and we probably do want to be quite careful about how we use it.

WATER

But that's enough about minerals, even though helium, by virtue of being a gas, is probably my favourite mineral: let's get back to the atmosphere. You might have noticed that there's sometimes quite a lot of water in the atmosphere, which can transfer to the Earth's surface in a process sometimes known to atmospheric scientists as "wet deposition", and to normal people as "rain". The amount of water in the atmosphere varies dramatically from time to time and place to place. In many parts of the world, rainfall is a precious commodity to be harvested, preserved, nurtured and used. In other places, including the damp little islands of the United Kingdom where I live, we generally have enough water to be going on with, so we are often a bit wasteful and even think of rainfall as a bit of a nuisance, especially when it interferes with our holidays or the cricket. And rainfall can be immensely destructive, causing floods and landslides, and knock-on effects such as enabling the spread of water-borne diseases and cutting communities off from essential supplies during a flood. Water in the atmosphere does have a role to play in some of the important atmospheric pollution processes, as we'll see. Chemical processes taking place within water droplets and on cloud surfaces can also be important – not least in stratospheric ozone chemistry – but we'll have to wait for Chapter 4 for that.

CARBON DIOXIDE

The next most prevalent substance in the atmosphere is carbon dioxide, CO_2. Jan Baptist van Helmont, the man who recognised the modern concept of a gas, was also the first to realise in the early seventeenth century that the gas that you get from burning wood or charcoal is also present in fermented food. This gas has to go somewhere, so it was reasonable to suppose that this gas would also be in the atmosphere. However, 400 years ago, there was no clear understanding of what this mysteriously prevalent gas actually was. Van Helmont named it "gas sylvestre" or "wild spirit". Great name, and in the process, van Helmont also invented the useful word "gas" by adapting the Greek word for "chaos". He carried out a famous experiment in which he grew a willow tree in a pot, weighed everything that went in and out, and showed that the tree had gained 164 pounds in weight over five years. It was slightly ironic that he didn't identify that the substance responsible for much of this weight increase was the same gas sylvestre on which he had carried out ground-breaking research. But his approach to scientific enquiry, based on careful weighing and control of the experimental conditions, was groundbreaking. More prosaically, carbon dioxide was called "fixed air" by Joseph Black in the following century. Prosaic is probably the right way to go: carbon dioxide is not a gas which engages in exciting reactions in the atmosphere, so "wild spirit" was perhaps putting it a bit strongly. However, carbon dioxide has had the last laugh, as we have come to understand more about its role in controlling the Earth's temperature. Even the nomenclature has come full circle, as we have moved from "climate change" to "climate chaos" – and the word "chaos" was the inspiration for van Helmont's neologism "gas" to describe carbon dioxide.

As a greenhouse gas, carbon dioxide has had a lot of press in the last twenty years – it's had much more attention than any of the more active shorter-lived air pollutants that I'm focusing on. Because of the role of carbon dioxide in climate change, putting policies in place to limit the level of carbon dioxide in the atmosphere is a key component of climate policy. For centuries before the industrial revolution at the start of the nineteenth century, the concentration of carbon dioxide in the atmosphere was 280 parts per million, or 0.028% of the atmosphere (nitrogen, oxygen, argon and water each make up about 1% or more of the atmosphere – then we jump down a long way to carbon dioxide, less than a thirtieth of one percent). Since we first acquired the habit of burning fossil fuels, concentrations have increased steadily until in 2016, the global average carbon dioxide concentration officially reached 400 parts per million, or 0.040%. That's quite an effect for one species to have on the atmosphere: without any help from anyone else, we humans have been able to up the level of carbon dioxide in the atmosphere by more than 40% in just a couple of hundred years. This level of carbon dioxide has serious implications for the global climate – without doubt, a huge, important and extensively written about topic. Important as climate change is, we're here to deal mainly with air pollution, which is a very different, and I think less familiar aspect of the atmosphere, so I'll do no more than dip a cautious toe into the science and politics of climate change.

CLIMATE CHANGE

Active as carbon dioxide is in climate terms, it is chemically pretty inert. That inertness is what gives carbon dioxide its long atmospheric lifetime, and enables it to make such an important contribution to long-term climate change. Much of the carbon dioxide that we discharge in the atmosphere today will still be there in a hundred years' time. That long lifetime is bad news for dealing with the effects of our activities on the global climate. And it also means that carbon dioxide is another chemically stable substance which doesn't play much of a role in the more frisky aspects of the atmosphere.

It's not as if the science of the atmosphere hasn't had a lot of attention recently. The role of the atmosphere in climate change has been the number one environmental story for well over twenty years. So much so that we humble air quality practitioners have perhaps struggled to make our voices heard over the huge investments in climate science, monitoring and policy development, which has been designed to understand the role of greenhouse gases and other aspects of the atmosphere in global climate change – and, of course, to take steps to do something about it.

For example, from 1993 to 2013, the US spent more than $165 billion on climate change.[20] That's a lot of dollars, and it's reassuring to reflect on how effective this spending has been in limiting atmospheric carbon dioxide to 350 parts per million (ppm, equivalent to 0.035%), following the advice of Professor James Hansen of NASA and Columbia University: *"If humanity wishes to preserve a planet similar to that on which civilization developed and to which life on Earth is adapted, paleoclimate evidence and ongoing climate change suggest that CO_2 will need to be reduced from [current levels] to at most 350 ppm."*[21] Except, of course, that the substantial and ongoing investment in research and impact mitigation mechanisms hasn't been at all effective in limiting increases in carbon dioxide levels in the atmosphere. In 2016, the global average carbon dioxide level passed the 400 ppm mark, and it's increasing by 2 ppm each year. Despite the major investments in research and control mechanisms made over several decades, the increase in carbon dioxide levels in the atmosphere is showing no signs of slowing down.

In some senses, dealing with our influence on the global climate is way simpler than dealing with air pollution impacts. Climate impacts are, to a large extent, directly linked to the presence of greenhouse gases in the atmosphere. The main culprit is carbon dioxide, responsible for about three quarters of annual greenhouse gas emissions, with smaller but still significant contributions from methane (16% of the total), nitrous oxide (6%) and refrigerant gases containing fluorine (2%).[22] The more of these substances there is in the atmosphere, the more we can expect the atmosphere to retain heat, resulting in increasing temperatures and unpredictable effects on climate and weather. From this perspective, climate science and mitigation policy is all about managing emissions. Once a greenhouse gas has been released into the atmosphere, we don't need to worry particularly about where in the atmosphere it is. The atmospheric lifetimes of the main greenhouse gases are long enough for them to mix effectively throughout the atmosphere, wherever they're released from, so that we are only interested in the quantity emitted (or the rate of emission) and the resultant global average concentration. Simples.

Air pollution, in contrast, is in many ways conceptually harder to get to grips with. As with greenhouse gases, we are, of course, concerned with emission quantities and emission rates, and the resulting concentrations in the atmosphere. In some cases, that's enough information to enable us to get on with managing the problem – for example, the United Nations Convention on Long-range Transboundary Air Pollution sets national limits on emissions of key pollutants from signatory states. The aim of this convention is "to improve air quality on the local, national and regional levels",[23] and the means by which it is achieved is to set limits on annual emissions at a national level. Also simples.

But to manage air quality in a complex environment, we need to look in more detail at the pattern of emissions, their dispersion in the atmosphere, physical and chemical conversion processes, and the resulting airborne concentrations. This kind of localised analysis gives us the information we need to understand and manage the impacts of air pollution. Without this kind of spatial detail, we've no way of knowing whether emission limits such as those set in the Convention on Long-range Transboundary Air Pollution will be good enough to ensure that we achieve air quality standards, and avoid excessive impacts due to high levels of air pollution. Or, conversely, maybe the limits are excessive, and go way beyond what's needed to avoid air quality impacts.

Of course, I'm dramatically oversimplifying climate science, in which complex interactions between the atmosphere, the oceans, the land, ecosystems and humans have to be investigated and understood. But my point is that air quality assessment requires consideration of the specifics of emissions to the atmosphere – the location, temperature, velocity and rate of release – in a way that assessment of greenhouse gas impacts doesn't. We are in the business of considering weather conditions, and the locations where people might be exposed to the released substances. We may need to understand background levels of pollutants in the atmosphere due to emissions from other sources. These factors are not directly relevant when dealing with greenhouse gases. Which isn't to say that dealing with greenhouse gases isn't complex enough. The complexities with greenhouse gases come from the need to establish unambiguous and accurate emission inventories, and interactions between the atmosphere and the planet, – perhaps more to the point – the urgent imperative to take action to manage (i.e. reduce) greenhouse gas emissions.

DEALING WITH CLIMATE CHANGE AND AIR QUALITY

And here we encounter a key cross-over between management of greenhouse gases, and air quality management. The most fundamental method of reducing both climate and air quality impacts is not to release so much pollution into the atmosphere. Any reduction in greenhouse gas emissions means slightly lower concentrations of greenhouse gases than would otherwise occur, and hence, marginally lower global temperature increases. Similarly, almost any reduction in emissions of air pollutants would result in lower levels of pollutants in the atmosphere than would otherwise occur.

This is all good so far as it goes, but of course the main reason we release pollution into the atmosphere is to deliver products and services which benefit and add value to our lives. Power stations generate electricity. Cars, buses, trains and aeroplanes give us mobility. Agriculture gives us food to eat. Domestic boilers and fires give us heat, hot water and enable us to cook. The list is endless. So of course it's not a simple matter of saying "create less pollution" if that means "produce less electricity" or "don't travel so much". Improving air quality has to take its place among a wide range of complementary and, in some cases, competing priorities.

Let's look at reducing how much pollution we emit in a bit more detail. How can we emit less? This can be achieved by <u>doing</u>

less, or by doing the same things more efficiently, or by applying technical fixes. Plus with air pollution, we can also sometimes reduce impacts by improving release conditions, by which I mean releasing pollutants faster, or at a higher temperature, or from a higher point. This can help to reduce local exposures to nearby sources of air pollutants, but wouldn't confer any particular benefit for greenhouse gases, however.

An example of reducing emissions by doing less is the reduction in the manufacturing industry in the UK over the past few decades. This has contributed to a significant reduction in emissions of substances associated with industrial activity – in particular, sulphur dioxide, as we'll see in the next chapter – along with looking at the question of whether this is a genuine reduction or just an outsourcing of pollution along with industrial activity to other countries. Sometimes, restrictions are placed on traffic or industrial/commercial activity at times of poor air quality, as a means of reducing impacts for a short period. A more positive example would be removing the need to carry out an activity – for example, enabling people to work from their homes, rather than travelling to and from an office. But in general, limiting activity is an unpopular means of reducing impacts. Simply requiring individuals or businesses to do less puts limits on the opportunities that people and businesses have to act as they wish. Short-term, possibly unpredictable limits on activity are difficult to enforce and have a disproportionately disruptive effect on everyday life and productivity. In overall terms, a normal economy encourages productive activity to take place, within defined boundaries – legal, ethical and environmental boundaries, for example. Placing restrictions on such activity, while it may be justified from an environmental or other perspective, inevitably requires the apparatus of regulation and enforcement.

So in principle, it's much better to secure air quality improvements not by restricting what people do, but by enabling people to do what they want to do more efficiently. And "efficiency" can cover a very wide range of options. Enabling people to travel by means other than private cars is often an effective way of reducing both greenhouse gas and air pollution emissions – even more so, because this kind of intervention often takes place in an urban setting, where levels of air pollutants may already be high. Alternatives to car transport may include public transport or improvements in cycling infrastructure. Even a policy shift to encourage car sharing may be beneficial in delivering reduced emissions of air pollutants and greenhouse gases. Low emissions zones, which place a charge on higher emitting vehicles, are being implemented to improve air quality and greenhouse gas emissions in cities around the world. And improvements in the efficiency of processes like domestic boilers, commercial space heating and electrical equipment will all result in less fuel being burnt, and hence, lower emissions of air pollutants and greenhouse gases.

So far, so lovely. By improving efficiency, or reducing activity, air quality and climate improvements can be delivered in harmony, giving a double bonus for investments in improved controls. This kind of change can also confer other benefits, such as reduced use of non-renewable materials, or lower noise levels. Wherever multiple benefits can be secured, it's usually a good basis for delivering air quality improvements.

DEALING WITH CLIMATE CHANGE, BUT NOT HELPING AIR QUALITY

Much of the focus of greenhouse gas reductions has been on the use of renewable energy. Some renewable technologies have the benefit of being zero or low emitting in terms of air pollution as well. Wind, solar and hydro power are genuinely zero emitting, and the replacement of conventional generation with these technologies would result in associated reductions in emissions of air pollutants at the point of generation. However, life becomes more complicated when we're considering renewable technologies which involve the combustion of wood, crops and biogenic waste materials. Sure, they are low emitting in terms of greenhouse gases. Biomass combustion does give rise to carbon dioxide emission, but because the carbon originates from biological sources rather than from fossil fuels, the carbon dioxide emitted from burning biological materials is considered not to make an overall contribution to greenhouse gas emissions. The carbon in wood originates from carbon dioxide in the atmosphere, and would have ended up as carbon dioxide again sooner or later, so burning the wood does not affect this overall cyclical process. The combustion process just makes sure that it's sooner rather than later. That means that we can (more or less) discount carbon dioxide from burning organic materials from contributing to greenhouse gas emissions.

However, these friendly, biomass energy sources, are definitely not zero emitting in terms of air pollution. From the wood burning stove in my sitting room to the latest wave of anaerobic digestors at farms and food factories, biomass combustion genuinely does emit air pollutants. And we can't make any distinction between biological sources and fossil sources in relation to air pollutants: pollutants emitted from biomass combustion are what they are. There's no trade-off with what might otherwise have happened. In many parts of the world, small-scale combustion of biomass such as wood and animal dung in homes is a major contributor to exposure to air pollution, both indoors and outdoors. As well as domestic cooking and heating, wood burning in small and medium sized plant to generate heat or electricity, or ideally both, is widespread throughout the world. While biomass combustion may be low-carbon, it isn't low-pollution. Handle with care.

DEALING WITH AIR QUALITY, BUT NOT CLIMATE CHANGE

As a result of the challenges posed by reducing activity and improving efficiency, much of the focus of air pollution control in the past twenty or thirty years has been on technical fixes. Low emissions combustion technologies in gas fired power stations, for example, designed to reduce the production of air pollutants.

Or catalytic convertors and particle traps on car exhausts, which remove air pollutants once they have been formed. Technical fixes and improvements are often great, but just as low-carbon energy isn't necessarily low-emission energy, low-emission solutions are not necessarily low-carbon.

The biggest air pollution solution to have been implemented in the past thirty years is the use of three-way catalytic convertors to reduce emissions of air pollution from petrol-engined road vehicles. In Europe, this has been enforced by increasingly demanding emissions standards for all vehicle types. In the case of petrol-engined cars, catalytic convertors have been needed to achieve these standards for cars manufactured since 1993. While the legislation doesn't specifically require catalysts, there is no other way to achieve the emissions limits. The next figure shows how effective catalyst technology and other improvements have been in reducing emissions from road traffic, despite ongoing increases in vehicle movements over that time. From 1970 to 1990, both the amount of road traffic and the amount of pollutants emitted from that traffic rose steadily. Since 1990, road traffic has continued its relentless rise, but emissions of air pollutants from traffic have gone completely the other way. Just look at the dramatic turnaround in carbon monoxide emissions from road traffic in 1990.

So catalytic convertors have been another success story, delivering substantial and sustained reductions in vehicle emissions of carbon monoxide, oxides of nitrogen and volatile organic compounds (universally referred to as "VOCs"). The number of vehicles on the road in 2014 was two and a half times higher than it was in 1970, but the air pollutant emissions from these vehicles in 2014 were all less than half what they were in 1970 (when I was three years old and moving to urban Essex, home of the Ford Capri, ready to experience the delights of London's finest exhaust pipes). Indeed, hydrocarbon emissions from road traffic in 2014 were just 5% of the 1970 emissions. That's a decoupling of pollution from economic growth that's worth shouting about.

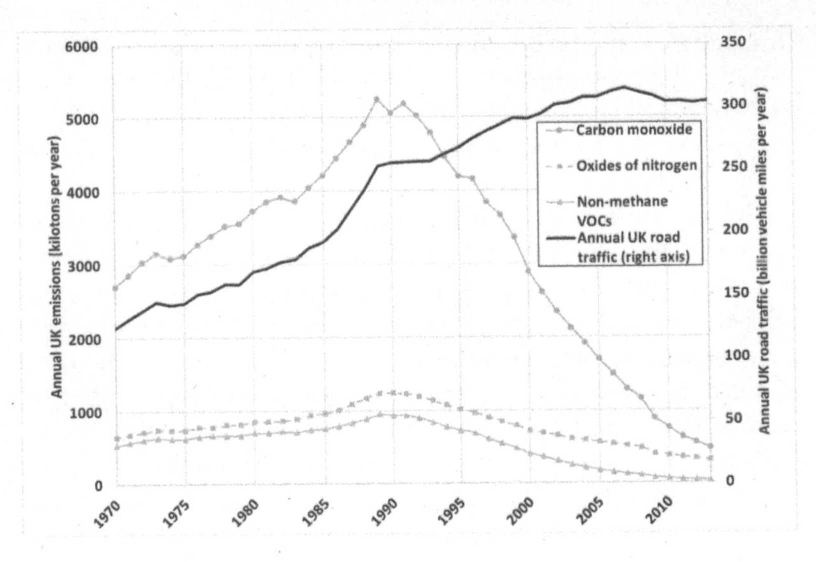

Vehicle mileage and emissions from road traffic in the UK, 1970 to 2013[24]

The key reactions within a catalytic convertor are the oxidation of carbon monoxide and hydrocarbons to form carbon dioxide and water, and the reduction of nitric oxide and nitrogen dioxide to form nitrogen – all substances which are inert from an air pollution point of view:

$$2CO + O_2 \rightarrow 2CO_2 + O$$
$$C_xH_y + (x + y/4)O_2 \rightarrow xCO_2 + (y/2)H_2O$$

(C_xH_y represents a hydrocarbon with x carbon atoms and y hydrogen atoms)

$$2NO \rightarrow N_2 + O_2$$
$$2NO_2 \rightarrow N_2 + 2O_2$$

A catalyst is a substance which increases the rate of a chemical reaction, without itself being changed during the reaction. In the case of the catalytic convertor, the catalyst is a finely divided metal with a high surface area – or in fact, usually two metals: one to speed up the oxidation of carbon monoxide and volatile organic compounds, and the other to speed up the reduction of NO and NO_2. The gases stick onto the catalyst surface, and the presence of the metal reduces the energy barrier for each reaction to take place.

A couple of paragraphs ago, I quietly mentioned that one of the side-effects of a catalytic convertor is to turn carbon monoxide and volatile organic compounds into carbon dioxide – which is, of course, a greenhouse gas. So while catalytic convertors have been sensationally good at reducing air pollution, they do result in a small additional penalty in terms of greenhouse gas emissions from cars. And it's a little bit worse than that, because sticking a catalytic convertor on the end of an exhaust pipe makes it harder for the engine to push the exhaust gases out. Because the engine has to work a bit harder, it is a bit less efficient than it would be if the catalytic convertor wasn't there at all. The slightly lower efficiency is another slight penalty in terms of greenhouse gas emissions. But on the whole, there is no doubt that catalytic convertors have been, literally, a life-saver.

Alongside the now universal use of catalytic convertors, vehicle engines have continued to become more efficient and better designed to reduce emissions of greenhouse gases and air pollutants. And there is more (by which I mean, less) to come, with a commitment to end the sale of petrol and diesel cars by 2040 in the UK and many other countries – more on this in Chapter 8. Lower emitting does not, of course, mean zero emitting, or at least, not yet, and road traffic continues to be a major source of greenhouse gases and air pollution. Worldwide, road traffic accounts for 14% of annual greenhouse gas emissions,[22] with most of the rest coming from industry, electricity and heat production, and agriculture. In London, road traffic accounted for about 30% of carbon dioxide emissions in 2013, and about half of particulate matter and oxides of nitrogen emissions (of which more in Chapter 3, 4 and most of the rest of the book.)[25] Looking forward to 2030, emissions of oxides of nitrogen are forecast to reduce further by almost 80% - great news if it happens, particularly in the context of the 76% reduction already achieved from the 1990 peak. Carbon dioxide and particulate matter from road vehicles are more intractable, however, with only slight improvements in emissions forecast between 2013 and 2030. One reason for this is that some of the particulate matter comes from sources other than exhaust emissions – tyre and brake wear, and vehicles stirring up dust from the road surface. These sources require a completely different approach to control and emissions reduction to the methods used to deal with exhaust emissions.

THE FUTURE FOR CLIMATE CHANGE AND AIR QUALITY IMPACTS

Some of the improvements needed to deliver further improvements in vehicle emissions of air pollutants will go hand-in-hand with reducing greenhouse gas emissions. More energy efficient vehicles will (probably) emit lower levels of all pollutants, both air pollutants and greenhouse gases. Other improvements might require a trade-off to be made between reducing greenhouse gas emissions, and reducing air pollutants, just like we did with catalytic convertors three decades ago.

And it's not just the design of the vehicle engine that matters. Air pollution depends on how, and how much, a vehicle is driven. Reducing the demand for use of road transport would be great for both the global climate and local air quality – but these environmental benefits have to take their place in the bigger picture. Vehicle use is a matter of consumer choice. The mobility provided by access to road vehicles is an essential component of quality of life for many people. Vehicles provide access to schools, jobs, friends, health care, leisure activities, all of which play a role in the overall health and well-being of a society. These benefits have to be considered alongside the downsides of road transport: as well as air pollution and greenhouse gas impacts, these downsides include noise, risk of accidents, community severance, and visual impacts.

Increasing the cost of using private transport is one approach to reducing vehicle use and associated impacts. This already happens with the Central London congestion charge, which at £11.50 a day has been comprehensively effective in dissuading me from driving into the centre of London at peak times to collect my son from university. A more widespread increase in the cost of motoring through the annual road tax or increased cost of fuel would, I suspect, be unpopular to the point of political suicide for anyone putting this forward as a serious policy. A significant increase in fuel duty would hit almost all of us pretty hard, and so would be something of a vote loser.

Although ... speaking as a completely unrepresentative member of the electorate (and also with no experience in fiscal policy), it seems to me that an increase in fuel tax is attractive in principle, because it would be easy to administer, it would hit the largest and most polluting vehicles hardest, it is correlated to how much you use your vehicle, and it could, if you like, be linked to the wider costs of vehicle pollution and greenhouse gas emissions. Of course, this approach would hit vehicle users across the board – both those who choose to use their cars, and those who don't have any option. It's interesting that the cost of a litre of petrol in the UK today in real terms is pretty much the same as it was in 1983,[26] and throughout that thirty-five year period, the cost has never varied by more than about 25%. So, despite a history of protests against the price of fuel, notably in 2000, 2005 and 2007, increasing fuel duty is an experiment that we've never really tried. Maybe it's worth a look?

Looking at how we drive our cars, rather than how much we drive them, recent initiatives have focused on speed limit controls. We've become increasingly familiar with variable speed controls on motorways, and anything which smooths out traffic speeds, avoids excessive acceleration and braking, and reduces congestion is going to be good for air pollution as well.

How about low speed zones? For example, most of Edinburgh's streets now have a twenty miles per hour limit applied. Twenty's plenty, or so they say, which is great for improving road safety, but doesn't driving at lower speeds result in less efficient engine operation, and higher emissions of air pollutants and greenhouse gases? Well, in fact, the evidence suggests that there isn't a big difference in emissions between a 20 miles per hour and a 30 mph zone, and in fact the 20 mph zone may result in lower emissions of particulate matter.[27] The important thing is to avoid speed bumps which encourage braking and acceleration. That's as bad for pollution as it is for my car's suspension.

Like many effective air quality and climate management measures, these controls on vehicle speeds are not primarily designed to deliver emissions improvements. The benefits for the atmosphere are a side-effect of a road safety or speed improvement measure. Well, there's nothing wrong with that, as long as the opportunities to secure air quality and climate benefits are understood and fully exploited along with the other benefits.

Maybe we will only finally nail the twin challenges of air pollution and climate change once we start to run out of fossil fuels – or perhaps a couple of hundred years after that, to allow time for carbon dioxide in the atmosphere to come back down. We have fossil fuel reserves for many decades to come, so the end of fossil fuels, with all the unknowns that will bring, isn't happening any time soon. But when it does come, what should we look forward to? Energy wars? A smooth transition to renewable energy technologies? A return to a stone age society? We still have the opportunity to make this transition a workable process, by inventing new technologies, rediscovering or adapting old ones, and perhaps changing our expectations in terms of the availability of cheap energy. Come back in 500 years, and find out how we've got on.

CHAPTER 3

A HAPPY ATMOSPHERE

Within A Thousand Kilometres

"I love a party with a happy atmosphere," sang Russ Abbot in 1984, in a rather less memorable song than The Police managed the previous year. What was it about the mid-1980s which spawned a generation of classic songs about air quality? Parties with happy atmospheres are, of course, great, and very much the natural habitat of the environmental scientist. Rather more importantly, we humans are lucky enough to live on a planet with a breathable atmosphere, if not a happy atmosphere, which, in the long run, is a whole lot more useful.

So at last we get on to the interesting bits of the atmosphere – interesting from the perspective of air quality, that is. In contrast with the global climate that we were thinking about in the previous chapter, air quality issues typically unfold over a regional or local area – say around a thousand kilometres or less. A region over which air quality issues develop is sometimes referred to as an "airshed", a term derived a bit awkwardly from the equivalent term "watershed" which we use to describe a ridge in the landscape dividing two river systems. You may have noticed that three dimensional air doesn't quite behave in the same way as water flowing over a land surface – not least, because air can, if pushed, go over hills, which I don't think rivers do – so I've never been completely convinced by the term airshed, but let's go with it, and start with the two big kids in the playground: airborne particulate matter, and nitrogen dioxide.

PARTICULATE MATTER

"Particulate matter" is a term which refers to solid particles in the atmosphere, and can cover all sorts of material. Particulate matter can include particles which are big enough to see with the naked eye – starting with grit and dust. These larger particles are important from the point of view of affecting visibility, and also from causing nuisance when the dust settles onto surfaces – buildings, cars or washing. There's more about assessing and dealing with dust in Chapter 7. But these larger particles don't have a direct effect on our health – partly because they tend to settle out onto the ground and other surfaces soon after they are released into the atmosphere, and mainly because they're too big for us to breathe them in.

We're usually more concerned with very fine particles – particles which are so small that you can't see them. These are the particles which are small enough to get past the very efficient filtration system provided by your mouth and nose, and enter the lung. We used to deal mainly with particles with a diameter of less than 10 microns, which used to be called "inhalable particulates", but are now universally referred to as "PM_{10}". To be strictly accurate, PM_{10} means "particles which pass through a size-selective inlet with a 50 % efficiency cut-off at 10 μm aerodynamic diameter."[28] If you've ever wondered what PM_{10} is, and why it's called PM_{10}, well, now you know, and I hope you're glad you asked. These particles are small enough to be inhaled through the mouth and nose, but it turns out that they are too big to enter our lungs. So attention has turned much more to still smaller particles referred to as "$PM_{2.5}$" or "PM_1" – that is, particles with a diameter of less than 2.5 microns, or 1 micron. These particles can get right down into your lungs when you respire (breathe in), which is why $PM_{2.5}$ is sometimes referred to as "respirable" particulate matter. And we sometimes study $PM_{0.1}$ – that is, particles with a diameter of less than 0.1 microns. A micron is a millionth of a metre, or (if you prefer), a thousandth of a millimetre. So you could line up 400 $PM_{2.5}$ particles between adjacent millimetre marks on a ruler, though it would need to be quite a dull afternoon for this to be an attractive option. The picture shows how big PM_{10} and $PM_{2.5}$ particles look compared to a human hair or a grain of sand from the beach.

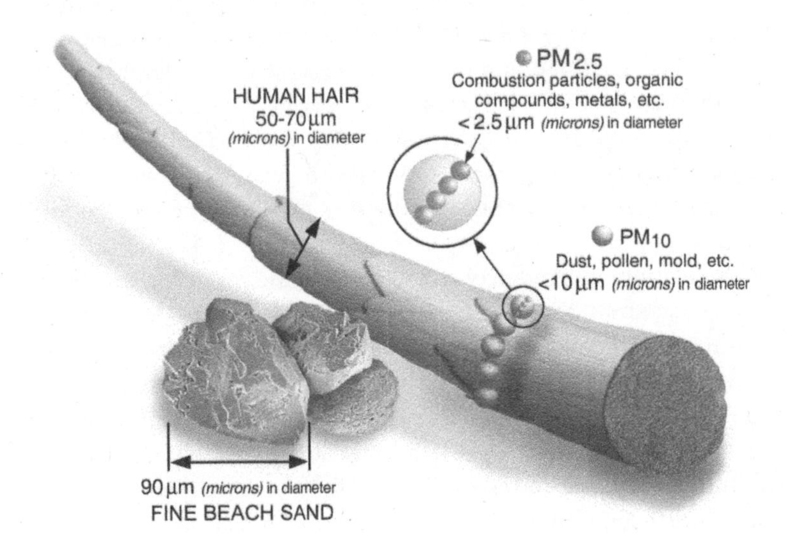

Illustration of PM_{10} and $PM_{2.5}$ particles[29]

We can't describe levels of particulate matter in percentage terms by volume as we can with gaseous pollutants. But we can describe levels as a mass concentration or in percentage terms by mass. One cubic metre of air at sea level weighs about 1.2 kilograms, or almost three pounds in old units. $PM_{2.5}$ levels in the atmosphere are typically around 10 micrograms per cubic metre ($\mu g/m^3$ – one microgram is a millionth of a gram). So $PM_{2.5}$ accounts for 0.0000008% of the mass of air. At this concentration, the amount of $PM_{2.5}$ in the air inside the Albert Hall would occupy about a quarter of a teaspoon, and weigh about the same as a small paper clip. It's not a lot. Even at the top end of $PM_{2.5}$ levels in the most polluted cities in the world, only about one twelve millionth of the mass of air that people breathe in comprises particulate matter sufficiently fine to get into their lungs. Yet this is the stuff responsible for seven million deaths per year worldwide. It's not surprising that this tiny quantity of material is hard to measure, harder to deal with, and harder still to demonstrate what you've done once you've done it.

OXIDES OF NITROGEN

Alongside particulate matter, the other pollutant which keeps a lot of air quality specialists busy is "oxides of nitrogen" – often abbreviated to "NO_x". NO_x is actually made up of two substances, nitric oxide (chemical formula: NO) and nitrogen dioxide (chemical formula: NO_2). These two substances are considered together, because there is an interchange between them in the atmosphere. In the presence of sunlight, nitrogen dioxide reacts to form nitric oxide. The reverse reaction also takes place, whereby nitric oxide is oxidised back to nitrogen dioxide. Most of the NO_x emitted from a combustion source like a car engine, domestic boiler or power station is in the form of nitric oxide, NO. As long as there are enough oxidants around, like ozone, the nitric oxide will react to form nitrogen dioxide. But in a large-scale plume, such as the emissions from a power station, the available oxidants are used up pretty quickly. Further oxidation of nitric oxide to nitrogen dioxide has to wait until the plume has dispersed so that clean air is mixed in with the power station flue gases. That can also happen in an urban area, where emissions from lots of road vehicles can use up the available ozone, with further oxidation having to wait for new oxidants to be formed through the action of sunlight. The reverse process is the splitting up of nitrogen dioxide to form nitric oxide. This only takes place in sunlight, so overnight, there is a shift towards a higher proportion of oxides of nitrogen being present as nitrogen dioxide because in the dark, the conversion between nitric oxide and nitrogen dioxide is a one-way street. During the day, the atmosphere is generally more active both physically (it's windier) and chemically (it's sunnier), which makes the conversion a two-way process. The net result is that nitric oxide makes up a higher proportion of oxides of nitrogen during the day. While the balance between nitric oxide and nitrogen dioxide shifts depending on all kinds of factors, the total amount of nitrogen present as nitric oxide and nitrogen dioxide added together remains roughly constant.

So what? Well, this interchange between nitric oxide and nitrogen dioxide does have some important consequences. The big issue is that nitrogen dioxide is harmful to our health, whereas nitric oxide is a lot less harmful. That's why there are air quality standards for nitrogen dioxide, but there isn't one for nitric oxide. When we're assessing air quality impacts, we often do a kind of two-stage process. Firstly, we calculate the levels of oxides of nitrogen due to emissions from all the sources under consideration. At this stage, we're not distinguishing between nitric oxide and nitrogen dioxide: we're lumping them all together. The second stage is to calculate, or estimate, or guesstimate, or guess the proportion of those oxides of nitrogen which are in the form of nitrogen oxide. So it can make quite a big difference if 10% of the oxides of nitrogen are in the form of more toxic nitrogen dioxide (as is typically the case in emissions from combustion sources) or 90% (as is often the case in the atmosphere overnight). One is nine times worse than the other. The combination of chemical and physical processes which affect the balance between the two components of NO_x make it more or less impossible to calculate the breakdown analytically in many circumstances, so we usually resort to guesstimating, while making sure that our guesstimates are on the safe side – that is, more likely to over-estimate nitrogen dioxide levels.

The balance between nitric oxide and nitrogen dioxide has also featured in our efforts to understand why air quality in cities wasn't improving as quickly as we thought it was going to. With new emissions controls on vehicles, we expected nitrogen dioxide levels to decline pretty rapidly through the 2000s. When they didn't, rather than knocking on the door of Volkswagen to ask nicely if they'd been fiddling the emissions tests, a lot of effort went in to looking at whether the latest generation of catalytic convertors affected the balance between nitric oxide and nitrogen dioxide in vehicle emissions. If a catalyst preferentially removes nitric oxide from the exhaust gases, this will raise the proportion of nitrogen dioxide in the exhaust gases. The UK's Air Quality Expert Group (a fine institution which very much does exactly what it says on the tin) produced a report in 2007[30] which highlighted the relatively high levels of nitrogen dioxide in emissions from diesel cars, and from buses fitted with particulate traps. So it's not just questionable emissions test results, important as those are: another reason why nitrogen dioxide levels haven't fallen as fast as we hoped they would is that vehicles achieve emission limits for NO_x by preferentially reducing emissions of nitric oxide rather than nitrogen dioxide. Once released, the two forms of NO_x interact so that you end up with the same equilibrium mix of nitrogen dioxide and nitric oxide after a while – but all that takes time, and particularly close to busy roads in urban areas, the result is higher levels of nitrogen dioxide than we expected.

There are other substances which might also be considered as "oxides of nitrogen", but these are less important in the atmospheric chemistry of NO and NO_2. For example, nitrous oxide (N_2O) tends to be formed and react in the atmosphere by different mechanisms to nitric oxide and nitrogen dioxide. It is emitted from agriculture, industry and combustion sources, and is stable in the atmosphere. Because of its stability, nitrous oxide is not an important contributor to atmospheric NO_x chemistry, but this very stability means that it is an important contributor to global warming. With a lifetime in the atmosphere of over 100 years, most of the nitrous oxide released today will still be there long after we're all gone. Now, that's a fairly dull description of nitrous oxide, which you probably recognised as laughing gas from the dentist's (if you're of a certain age) or maternity suite (if you're of a certain gender) or the streets of Kavos (if you're of a different age). As a fifty-year-old gent, thank goodness I'm too young, too male and too old respectively to have experienced laughing gas in any of these ways. Can't say I'm desperate to try it myself, having seen the effects that it has on people inhaling it for fun. Also known as hippy crack or just "nos", it gives you a short-lived and pretty fuzzy kick, often with a fit of the giggles. I guess that's OK so far as it goes, but users are briefly out of control and lose perception of their surroundings. Not to mention leaving a litter of used balloons and capsules behind them, and making a small, personal and very direct contribution to climate change.

Because of the conservation between nitric oxide and nitrogen dioxide, policies and initiatives relating to air pollution emissions (that is, what goes into the atmosphere) typically focus on total NO_x. That makes sense, because NO_x is easy to measure and control. However, policy and initiatives relating to ambient air quality (such as air quality standards set to protect public health) normally address nitrogen dioxide because it is much the more harmful to health of the two substances. That makes sense too. The problems have started to mount up when we were not paying enough attention to the difference between nitrogen dioxide and NO_x. The relationship between NO_x and NO_2 changes not only from place to place, from day to night and from summer to winter, but also more subtly in the long term as sources of oxides of nitrogen evolve. The good news is that with better information on the breakdown of vehicle emissions between nitric oxide and nitrogen dioxide, we can start to get a better grip on current sources and future controls on NO_x.

Oxides of nitrogen are typically present in the atmosphere at similar concentrations to $PM_{2.5}$ – so again, only about 0.000001% of the atmosphere. And once again, this tiny proportion is enough to contribute to respiratory ill-health, as well as legal headaches. Legal headaches? Yes – because in Europe, for example, there is a mandatory air quality standard for nitrogen dioxide of 40 micrograms per cubic metre ($\mu g/m^3$). The UK and a number of other EU Member States are facing legal challenges because of a failure to achieve this air quality standard everywhere within the country, as quickly as possible. The air quality standard for nitrogen dioxide is achieved throughout most of Europe, including most of the UK, but it is exceeded in some limited areas – mainly the centres of large cities and conurbations. For example, at the time of writing, the UK has been dragged back to the European Court of Justice with regard to breaches of the air quality standard for nitrogen dioxide in sixteen areas within nine conurbations.[31] London is, not surprisingly, one of the areas affected by high nitrogen dioxide levels, along with Glasgow, Teesside, Greater Manchester, West Yorkshire, Kingston upon Hull, the Potteries, the West Midlands and Southampton. Alongside this, the UK government has faced, and lost, separate legal proceedings, also linked to nitrogen dioxide levels, and the need to comply with the standards in the shortest possible time. It's taking a lot of work to understand and take steps to improve air quality in these areas in the shortest possible time, as required, and all because nitrogen dioxide weighs in at slightly more than 0.000003% of the atmosphere in these areas. One way out of this legal difficulty might be, say, to exit the European Union. This feels kind of like a big step, and while it might enable us to extricate ourselves from some of these legal proceedings, I can't see it helping to actually deliver improvements in air quality in these cities. Meanwhile, we in the UK are not alone: the European Commission is threatening legal action against eleven other member states, and ominously says, "Action against other Member States may follow."[32] You have been warned.

In the context of these widespread problems with urban nitrogen dioxide throughout Europe, one might expect similar issues to arise in the United States. But no, there are no "non-attainment zones" for nitrogen dioxide anywhere in the US. The reason for this is pretty simple: the annual mean air quality standard for nitrogen dioxide in Europe is 40 $\mu g/m^3$, whereas in the US it's a much less demanding 102 $\mu g/m^3$. The US National Ambient Air Quality Standard wouldn't have been exceeded anywhere in the UK except for a few roadside locations. The US standard was set in 1971, and although it's been reviewed a few times since then, the standard value has not changed. In contrast, the US air quality standard for $PM_{2.5}$ was set in 2012 at a pretty demanding 12 $\mu g/m^3$, close to the World Health Organization's guideline of 10 $\mu g/m^3$. As a result, there are nine non-attainment zones for $PM_{2.5}$ in the US, mostly in California. Looks like we in Europe need to catch up with the US on $PM_{2.5}$ standards. In a global context, however, things aren't too bad: the World Health Organization's Global Health Observatory publishes a biennial data compilation[33] which suggests that average $PM_{2.5}$ levels are 12 $\mu g/m^3$ or below in just 28 of the 184 countries for which data are available (the UK just failing to make the cut, at 12.1 $\mu g/m^3$). The countries to steer clear of if you can, with average urban $PM_{2.5}$ levels of over 100 $\mu g/m^3$, are Qatar and Saudi Arabia. There are a further fifteen countries averaging over 50 $\mu g/m^3$, which altogether account for almost half the world's population (in increasing order of urban $PM_{2.5}$: Myanmar, China, Uganda, Niger, Pakistan, Bahrain, Libya, India, United Arab Emirates, Nepal, Mauritania, Cameroon, Kuwait, Bangladesh, Egypt, Qatar and Saudi Arabia). That gives us a clue as to why $PM_{2.5}$ has such a debilitating effect on the health of people worldwide: half of us are living in countries where urban $PM_{2.5}$ levels are between four and ten times as high as the US air quality standard. It's not surprising there's a problem.

CARBON MONOXIDE

If I had been writing this book a few years ago, I would have been talking in similar terms about carbon monoxide and sulphur dioxide. But times have moved on. With the introduction of catalytic convertors on road vehicles, carbon monoxide levels have dropped dramatically. Levels of carbon monoxide in the atmosphere remain around ten times higher than levels of nitrogen dioxide – but the air quality standard for safe levels of carbon monoxide is fifty times higher than safe levels of nitrogen dioxide. So, as long as short-term peak nitrogen dioxide levels are there or thereabouts (that is, complying with the air quality standard for hourly mean nitrogen dioxide concentrations), which they invariably are in the UK, carbon monoxide levels are not normally anything to worry about.

SULPHUR DIOXIDE

Sulphur dioxide is another success story – at least, in some parts of the world. The figure on the following page shows how measured levels of sulphur dioxide in the UK have changed since 1990.

This looks like a pretty remarkable decrease in sulphur dioxide levels over a twenty-five year period, but in fact, it's just the tail end of a longer-lasting decrease. The figure on page 59 shows the same measurements, but with levels measured at the now-closed Central London site between 1974 and 1989 added in.

Now, that is a really dramatic improvement in air quality – a genuine success story. Sulphur dioxide was a major component of the famous London smogs of the mid-1950s. When I first became interested in air pollution in the late 1980s, acid rain resulting from sulphur dioxide emissions was one of the hottest topics in town. Nowadays, we are no longer talking about acid rain, and airborne sulphur dioxide doesn't get much of a mention as an important air pollutant, at least in the UK. One of the key factors in this remarkable improvement is the decline of the manufacturing industry in the UK and elsewhere in Western Europe.

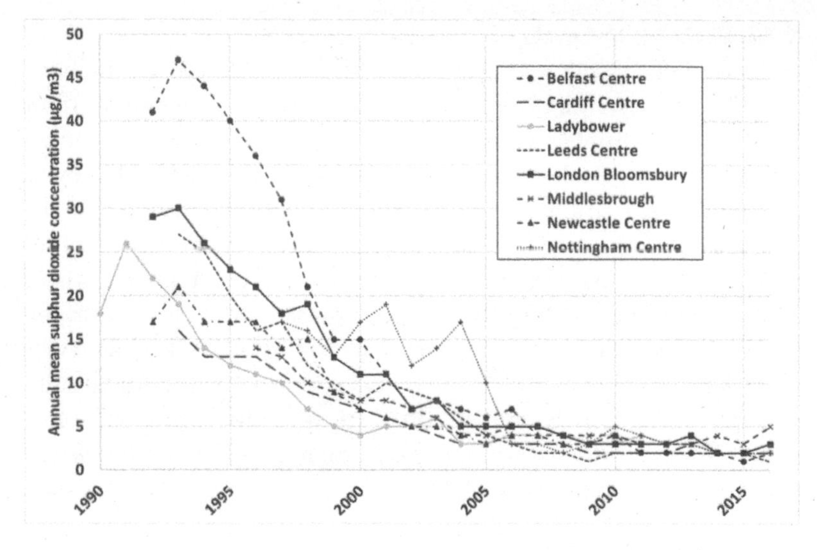

Measured levels of sulphur dioxide, 1990 to 2016[34]

Looking at the global picture for sulphur dioxide, there is one source which has shown consistent, strong growth for the past twenty years while every other sector is reducing or at worst staying more or less constant, and that is international shipping. Between 1990 and 2011, shipping emissions doubled, so that this category accounted for about one seventh of global sulphur dioxide emissions by 2011. Of course, a lot of this sulphur dioxide is emitted in the middle of the ocean, so it tends not to have a direct effect on people, other than passengers on cruise liners, who are as deserving of protection from environmental pollution as any other dispossessed and stateless minority.

As well as a reduction in industrial activity as we in the UK have moved to a more service-focused economy, the success story in reducing sulphur dioxide owes much to effective regulatory intervention through the Clean Air Acts, together with wider scale changes in sources of air pollutants. Not only have a lot of

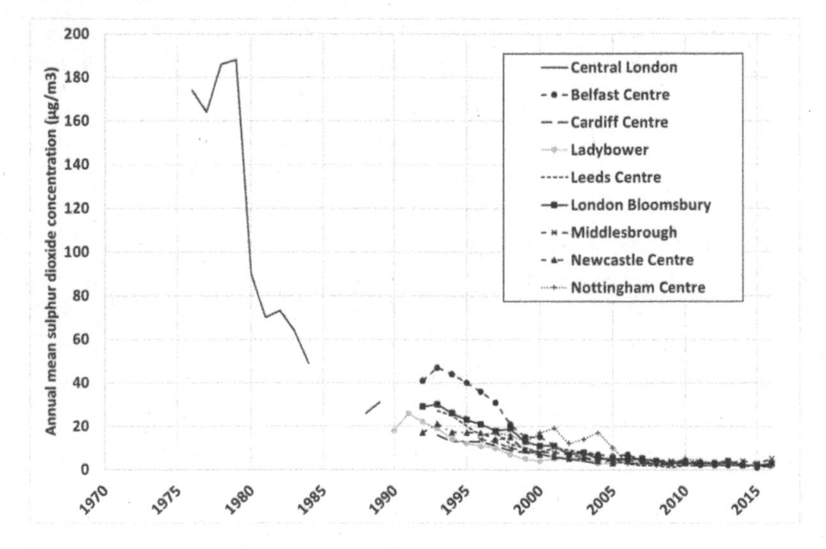

Measured levels of sulphur dioxide, 1970 to 2016[35]

industries scaled back dramatically, but we have also moved away from burning coal and oil in factories, power stations and in the home, with a focus on cleaner fuels. Because of these changes, coal consumption in the UK fell by 83% over the sixty years between 1955 and 2015. Most of the coal consumed in the UK nowadays is burnt in power stations, and any power station using coal has to operate a flue gas desulphurisation system. This usually takes the form of a fine spray mist or alkaline powder which absorbs the sulphur dioxide and is removed from the flue gases before they are discharged to the atmosphere. It can be over 90% effective in removing sulphur dioxide from flue gases. As a result, while coal consumption for electricity generation halved between 1970 and 2014, emissions of sulphur dioxide from coal burning in power stations fell by 95%. Sulphur dioxide removal isn't particularly new or innovative technology, but there was no specific regulatory or legislative driver to reduce emissions until national ceilings were imposed at a European level, and these were enforced on power station operators through the licensing and permitting system. It's amazing what you can do when you have to.

The improvement in sulphur dioxide levels in the UK and other first world economies hasn't happened in isolation: as we've reduced the intensity of manufacturing in Europe, this has been more than matched by a rise in industrial activity in other countries, notably India and China along with other Asian economies. So in one sense, we in the UK have stopped exporting our pollution directly and conveniently to our neighbours downwind in Scandinavia and central Europe, and started exporting it indirectly to other parts of the world. And that's been reflected in increasing pollution levels in growing economies. For example, whether it's due to manufacturing goods for export to the UK or elsewhere, or for some other reason, levels of sulphur dioxide measured in India doubled between 2005 and 2015.[35] In 2014, India overtook the USA to become the second-highest sulphur dioxide emitter in the world, after China.

This is largely down to our old adversary coal, in the reverse process to the de-coaling of first world economies. One of the most attractive features of coal for industrial process operators is that it's pretty cheap. And there are also substantial reserves in China (which produces almost half the world's coal), India and Indonesia. Consequently, much of industrial growth in economies like India and China has been fuelled by coal, including the use of coal for electricity generation. This results in an increase in sulphur dioxide emissions to the atmosphere, particularly in countries where (as was the case in the UK during past decades) there is no strong incentive to reduce these emissions using flue gas desulphurisation technologies.

Sulphur dioxide emissions from India were on a steady and dramatic upward trend from 1990 right through to 2012, and possibly beyond. In contrast, China (with total emissions around three times higher than India's) applied emissions limits from the

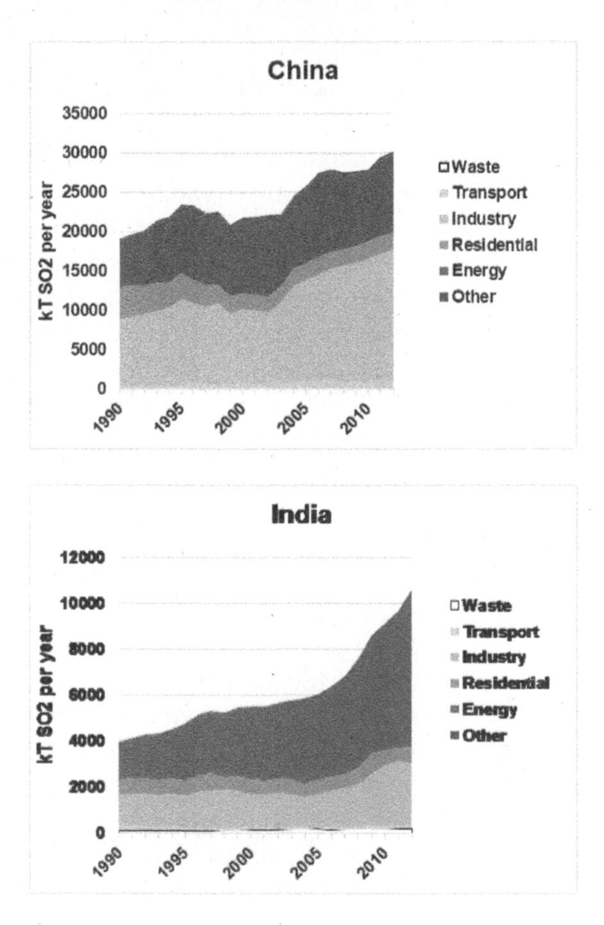

Trends in sulphur dioxide emissions from China and India 1990 to 2010[36]

early 2000s. This was followed by a programme of implementing flue gas desulphurisation, resulting in a reduction in estimated sulphur dioxide emissions from 2005 onwards, while coal consumption continued to rise. The rise in coal consumption in China and associated carbon dioxide emissions and other environmental impacts are well documented. But it has at least been possible to decouple this from an ongoing increase in sulphur dioxide emissions, and start to bring these down. There is a long way to go, however: annual average concentrations of sulphur dioxide in Beijing remain up to 50 $\mu g/m^3$, high enough to exceed China's Grade I air quality standard for annual mean concentrations, and probably high enough to result in exceedances of the national standards over shorter averaging periods, too.

To deliver ongoing improvements in sulphur dioxide levels worldwide will require the kind of measures and changes which have taken place in Western Europe since the 1950s. Our experience is that reducing sulphur dioxide emissions requires regulatory controls on a wide range of sources, from small-scale industrial installations and domestic fires, through to large power stations. In many ways, controlling sulphur dioxide emissions from large industrial sites and power stations is the most straightforward part of the picture. They're very visible sources, with one or two big chimneys where emission limits can be applied, and abatement installed. It's much harder to get to grips with small-scale sources, such as small brick kilns which can be set up almost anywhere with minimal controls on combustion and emissions. Widespread burning of solid fuels in the home can also be important.

Dealing with these small-scale sources requires a combination of carrot and stick. The stick in this case is restrictions on the use of solid fuels, backed up by inspection and enforcement. The carrot includes incentives to use alternative low-sulphur fuels such as liquefied natural gas, perhaps enough to make the cleaner alternatives cost-effective compared with solid fuels. More fundamentally, it's important to make sure that there are alternative fuels available to enable communities and industrial process operators to meet their needs for heating without the use of high-sulphur coal and oil. This is perhaps the real challenge, particularly in rural areas. In the UK, we dealt with this by building the national gas transmission and distribution systems from the 1960s onwards. Nowadays, 84% of homes in the UK have access to piped natural gas. Natural gas is such a great fuel – none of that tedious, dusty carrying around of coal and getting rid of ash to worry about – that domestic users lap it up whenever it's available. It's an expensive job providing piped natural gas to domestic and business consumers, and it is of course a fossil fuel with finite reserves. But maybe targeted provision of natural gas can be part of an effective strategy to reduce the use of solid fuels in areas experiencing high levels of sulphur dioxide. There are plenty of other options to consider – improving energy efficiency is important, of course, along with zero-emissions energy sources including solar, geothermal, hydro and ground source heat pumps. Biomass-based sources such as anaerobic digestion to produce biogas, or combustion of wood, waste or purpose-grown crops may also have a role to play, although these sources have their own pollution impacts which need to be properly managed.

CLOSER TO HOME

So what a mash-up the atmosphere is. Sure, there's a lot of argon and nitrogen sitting around not doing very much. Mixed up with the inert gases, there is quite a lot of oxygen doing nothing less than giving us life with every breath we take. And speaking of life-giving, there's often a load of water up there in all its many forms – the very tangible rain, snow, sleet, hail, mist and fog, as well as invisible water vapour. But the atmosphere that giveth also taketh away, through a few greenhouse gases, and extremely small quantities of stuff that seems to have a quite remarkable effect on our health, on ecosystems, on visibility, and which can sometimes be quite a nuisance, as it gets closer and closer to home. We'll have a look at these substances in the next few chapters.

CHAPTER 4

OZONE: GLOBALLY HELPFUL, REGIONALLY HARMFUL

Still Within A Thousand Kilometres

WHAT'S OZONE?

Oxygen is the element that gives us life. It's reactive, exciting and dynamic. Oxygen is mad, bad and dangerous to know.

What oxygen is good at is taking chemicals made up mainly of hydrogen and carbon, and converting them to chemicals containing hydrogen, carbon and oxygen. If this process goes all the way, you end up with carbon dioxide and di-hydrogen monoxide – and if you're wondering what di-hydrogen monoxide is, it's H_2O, perhaps more commonly known as "water". Conveniently, these chemical transformations result in a net excess of energy. This process is, roughly, what we refer to as "life" – at least, for those of us who get our kicks from breathing rather than photosynthesis. In other circumstances, we can equally refer to this process as "fire" – a more dramatic way of generating heat and light from the reaction between oxygen and hydrocarbons. And if it's sunlight rather than combustion driving the oxidation of carbon and hydrogen, we might call it "photochemistry". But however they happen, the reactions between hydrogen, carbon and oxygen are, in every sense, elemental.

This energy-giving role makes oxygen a vital element for our very existence. But the same characteristics that make oxygen ideal for sustaining life bring an element of danger. We need oxygen to react with the starches, sugars and proteins in our diet to generate the energy that we need to live. But oxygen is so reactive that it can also damage the cells which are the building-blocks of our bodies. These cells that make up our bodies are built out of carbon, hydrogen and oxygen, and so they too can be susceptible to damage from oxygen.

In the atmosphere, almost all oxygen is present in a pretty stable form, O_2. The number 2 in this chemical formula represents two oxygen atoms which are chemically bound to each other. Only one thing in the atmosphere is energetic enough to break up this stable relationship: enter sunlight. High up in the atmosphere, the energy from sunlight in the ultraviolet part of the spectrum is energetic enough to react with di-oxygen (O_2) molecules to form two oxygen atoms.

$$O_2 + Sunlight \rightarrow O + O$$

The stronger the sunlight is, the more rapidly this reaction can go. But the forces which bind those two oxygen atoms together are very strong, so it takes place only slowly even when oxygen is sunbathing in a bikini on a summer afternoon. And once those strong bonds have been broken, the chemical driving forces which made them so strong are still at work, making those single oxygen atoms very, very reactive. Because they are so unstable and ready to react with almost anything, they are known as "radicals". The oxygen atoms formed in this reaction mainly combine with another oxygen molecule to form a strange and interesting chemical called ozone.

$$O + O_2 \rightarrow O_3 + Heat$$

Ozone consists of three oxygen atoms in a triangular form, looking something like this:

An ozone molecule

Ozone is a pretty unstable molecule. It takes a lot of energy to hold the three oxygen atoms in the slightly awkward triangular form, held together by two lots of one and a half bonds. Chemists use the term "resonant structure" to describe this slightly odd arrangement, whereby ozone needs a total of three covalent bonds to achieve complete sets of electrons for each of the three oxygen atoms. But it's not a particularly happy solution, which leaves the ozone molecule with a bit of a positive charge on the middle oxygen atom and a negative charge on either side. This makes ozone a reactive substance, ready to react with a lot of other substances, giving up an oxygen atom and reverting to the much more stable oxygen – two oxygen atoms held together by a double bond.

OZONE IN THE STRATOSPHERE

Because of its complex chemistry, ozone operates on a regional scale. It takes a while for the chemical and physical processes which affect ozone to play out, so ozone evolves over distances of hundreds or thousands of kilometres, and crucially behaves very differently at different heights in the atmosphere. The chemistry of ozone high up in the atmosphere was first unravelled by Sydney Chapman in the 1930s. Chapman trained as an engineer and mathematician in Manchester and Cambridge, taking advantage of the exchange of people and ideas between those universities – the mathematicians John Littlewood and, later, Alan Turing also travelled in both directions. After graduating from Cambridge, Chapman began work at the Royal Observatory in Greenwich, but found that he wasn't inspired by astronomy. He soon found his vocation as a pioneer and developer of solar terrestrial physics: that is, the interactions between the sun's radiation and the Earth's atmosphere and climate, and who am I to disagree that atmospheric science is a better career option than astronomy? In a landmark paper published in 1930, Chapman proposed a small set of chemical and photochemical reactions which are the starting point for our current understanding of the processes which govern the level of ozone in the atmosphere.

Ozone can itself be split by ultraviolet sunlight, re-forming an oxygen molecule (O_2) and an oxygen atom (O).

$$O_3 + UV\ light \rightarrow O_2 + O$$

The oxygen atom then quickly reacts with an oxygen molecule, and we're back with ozone again where we started. Well, nearly where we started. The net effect of this process is the absorption of ultraviolet light, and the release of a little bit of heat. This cycling round and round between ozone and oxygen atoms continues until an oxygen atom happens to bump into an ozone molecule or an oxygen atom, rather than an oxygen atom. Or, theoretically, two ozone atoms collide and react. All these reactions result in a more stable oxygen molecule, and so the reactive oxygen atoms and ozone molecules are out of the game.

$$O + O \rightarrow O_2$$
$$O + O_3 \rightarrow O_2 + O_2$$
$$O_3 + O_3 \rightarrow O_2 + O_2 + O_2$$

It turns out that the levels of oxygen atoms present in the atmosphere are so small that we can neglect the reaction between two oxygen atoms. And the reaction rate of two ozone molecules is pretty slow too, so we can also neglect this process. Some straightforward-ish maths enabled Chapman to outline how ozone concentrations would vary depending on the time of day, time of year and latitude. This set the framework for stratospheric ozone chemistry: all he needed was some decent data.

All these reactions happen up in the stratosphere, where the rate at which ultraviolet sunlight forms ozone, and the rate at which it is removed, balance out, enabling detectable levels of ozone to be present. Higher up, and the rate of removal by sunlight or reaction with other radicals is too great for ozone to survive. Lower down, and there isn't enough ultraviolet light to form ozone in the first place, because it's all been absorbed by stratospheric ozone at higher altitudes. That's why we end up with an ozone layer at the bottom of the stratosphere. The amount of ozone present in the stratosphere is determined by the balance between the formation, cycling, and removal of reactive oxygen atoms.

OZONE AND ULTRAVIOLET LIGHT

We've already encountered the heat produced when an oxygen atom and an oxygen molecule combine to form an ozone molecule. The heat generated by this process is what causes the tropopause – the abrupt change in temperature profile which marks the boundary between the troposphere and the stratosphere. It's no coincidence that the ozone layer crops up at the point where the troposphere stops and the stratosphere starts – it's the same process responsible for forming ozone and changing the atmosphere's temperature profile. Another very handy by-product of this process is the absorption of ultraviolet light by ozone molecules – the reaction I showed above:

$$O_3 + UV\ light \rightarrow O_2 + O$$

A quick digression for a minute. Light radiation is carried by fundamental particles called photons. Light has a characteristic intensity and a characteristic range of wavelengths. The intensity of the light is a function of how many photons there are – more photons, greater intensity. Each photon has a single wavelength, and this determines how much energy each photon carries. Shorter wavelength, more energy in each photon. So each photon of short-wavelength violet light carries a lot more energy than the longer wavelength photons of red light. Light which has too short a wavelength for us to see is called "ultra-violet" light. And the energy of each photon of ultraviolet light gets so great that it starts posing a threat to us. This is where the layer of ozone twelve kilometres above our heads does a vital job. The sun radiates energy across a wide range of wavelengths. This includes visible radiation of course, but also longer wavelength infrared radiation (which we feel as heat) and shorter wavelength ultraviolet radiation. We can't normally feel or detect UV radiation, but the photons of UV light are so packed with energy that they can start to interfere with our bodies' cells, and that's why prolonged exposure to sunlight can accelerate the aging process, and increase the risk of skin cancer. Each photon of ultraviolet light from the sun that is absorbed by ozone in the stratosphere is one less photon that might make its way down to your back garden where, who knows, you might be sunbathing. And there is enough ozone in the stratospheric ozone layer to absorb almost all the dangerous ultraviolet light, and allow us to sunbathe with confidence, as long as we've got some factor 30 on. But that layer of ozone is fragile: if all the ozone in the stratosphere was brought to ground level, it would make a layer in the atmosphere just three millimetres thick.[37] What if it wasn't there?

MEASURING OZONE

That wasn't a problem for a long time: ozone in the stratosphere went quietly about its business, protecting us from ultraviolet radiation from the sun, for thousands of years. In 1957, scientists from the British Antarctic Survey at Halley Bay began monitoring stratospheric ozone using a very basic Dobson spectrophotometer, so basic that it had to be wrapped in a duvet before it would work properly[38] (the same would probably be true of me if I ever get to the Antarctic). The Dobson meter chugged away for fifteen years, recording ozone levels in the atmosphere directly above the research station, without registering anything particularly interesting. Then in the 1970s, Paul Crutzen, Mario Molina and Sherwood Rowland developed our understanding of how the stratospheric ozone layer is formed, building on the Chapman mechanism, and raised concerns about the effect that man-made chemicals, especially chlorofluorocarbons (CFCs), could have on the ozone layer. These concerns surrounded the stability of CFCs. Stability was one of the features that made this group of chemicals attractive as refrigerants, but it also means that they have a very long atmospheric lifetime. Long enough to circulate through the lower atmosphere and upwards into the stratosphere. Crutzen and colleagues realised that they wouldn't break down until they were exposed to high energy ultraviolet light. And that wouldn't happen until the CFCs reached the same height as the stratospheric ozone layer, and started to lose the protection from ultraviolet light that the ozone layer gives to both CFCs and sunbathers at lower altitudes. Once in the stratosphere, there would be enough high energy ultraviolet light to start the process of breaking down CFCs. So releasing CFCs into the atmosphere looked like it would be a way of delivering highly reactive chlorine and fluorine atoms into the stratosphere, right among the very useful ozone layer. What would these atoms do? The worry was that they would catalyse the removal of ozone from the atmosphere, via a very efficient chemical process:

$$Cl + O_3 \rightarrow ClO + O_2$$
$$ClO + O_3 \rightarrow Cl + 2O_2$$

The net result is conversion of two molecules of ozone to three molecules of oxygen, with the return of the chlorine atom to go through the process again, and again, and again.

Twenty years later, this research would result in a Nobel Prize for Crutzen, Molina and Sherwood. But in the 1970s and early 1980s, there was no evidence of any problems with the ozone layer, either from the single instrument at the British Antarctic Survey research station at Halley Bay, or from the more extensive and high-tech satellite-based measurements being made by NASA. And the theoretical models indicated that there weren't yet enough CFCs in the atmosphere to cause a major problem for the ozone layer.

Then in the Antarctic winter of 1982, Joe Farman of the British Antarctic Survey noticed a sudden drop in the ozone level above the Halley Bay instrument. The Dobson meter made a measurement at a single point, and the apparent drop in ozone level at Halley Bay was not corroborated by any other data, so Farman thought it was probably a malfunction and ordered a replacement instrument. The next year, the drop in ozone levels happened again, down to about half the normal level. By now, having been reproduced on two separate instruments, it was looking like a genuine observation, but was it a widespread problem, or just limited to the atmosphere above the Halley Bay research station? In 1984, Farman and colleagues looked further afield, and found the same thing. They couldn't sit on the findings any longer and so, against some considerable opposition, Joe Farman and his colleagues Brian Gardiner and (in his own words, "general dogsbody") Jonathan Shanklin[39] published their findings in the May 1985 edition of *Nature*.[40] The stakes were high: if they were right, there were long-term implications for the future use of refrigerants and possibly an increased risk of cancer for people living under the ozone hole – and they had no idea how widespread the problem might be. If they were wrong (as seemed entirely possible, because no one else was reporting low ozone levels) it would be a severe embarrassment for the British Antarctic Survey and their sponsors.

This publication caused a major stir: it confirmed what Crutzen, Molina and Sherwood had in principle predicted, and by linking the measured decline in ozone levels to stratospheric NO_x and chlorine levels, clearly implicated human activity in emitting chlorine-containing CFCs into the stratosphere.

NASA had two ozone monitoring instruments on the Nimbus-7 Satellite, which had been launched in 1978 with a design life of one year, but which were still operating in 1985. The NASA team went back to look again at data from its Total Ozone Mapping Spectrometer (TOMS) instrument on Nimbus-7. They found that the instrument had been set up to reject any data which deviated too far from the expected amount of ozone.[41] At that time, the influence of CFCs on total ozone levels was expected to be pretty small. Consequently, the measurements which should have rung alarm bells as the ozone hole grew larger year by year were not reported by the instrument, and no one was checking on the extent of data rejection by the instrument's software or investigating the reasons for data being discarded.

THE STRATOSPHERIC OZONE HOLE

NASA's data confirmed the single point measurements from the British Antarctic Survey, and showed that the ozone hole was widespread, and was mostly over Antarctica. During the Antarctic winter, the air over the South Pole rotates around the pole, forming a stable air mass known as the "polar stratospheric vortex." This results in an isolated, stable air mass, within which the ozone depletion cycle operates efficiently in the gas phase and on the surfaces of polar stratospheric clouds. These icy and acidic clouds are made up of tiny particles, and as a result, have a large surface area on which chemical reactions can take place, ending up with the removal of ozone from the stratosphere. This explains why the ozone hole is deeper than anyone had predicted, and also why it is conveniently centred over an uninhabited part of the world. If you had to have an ozone hole anywhere, you'd probably pick Antarctica, although in fact, the phenomenon of ozone holes is not limited to the Antarctic. The ozone layer is affected all over the world, but because the conditions in more densely populated parts of the world are, fortunately, less favourable for these reactions to progress, stratospheric ozone depletion is much less dramatic away from the Antarctic.

The British Antarctic Survey and NASA observations were a remarkable confirmation of the theoretical predictions of atmospheric scientists, which gained worldwide publicity. The story brought together matters of genuine concern (increased risk of cancer, as well as harm to the pristine Antarctic environment from human activity) together with familiar sources of the problem (fridges and aerosol cans) and unfamiliar concepts of atmospheric science ("Ozone? Isn't that what smells so great at the seaside?").

Action to deal with the problem followed remarkably rapidly. It was only a couple of years later that the Montreal Protocol on "Substances that Deplete the Ozone Layer" was signed. This has been followed by action to ban and replace the substances which caused the problem, and in some cases to ban the substances that replaced the substances that were banned, when they turned out to be almost as bad as the substances they were replacing. It was possible to take such swift action because of a number of factors. Firstly, the science was pretty clear and unambiguous (much like the evidence for the human influence on climate change today). Secondly, the possible effects on health were severe and easily recognisable. And thirdly, the measures needed to deal with the problem were down to product developers and manufacturers. The aerosol spray and refrigeration industries were able to identify and implement replacement chemicals pretty quickly and at reasonable cost. Consequently, there was no need for potentially unpopular large-scale changes in behaviour or cost increases for consumers.

So all we need to do now is wait for seventy years or so for the chlorine and fluorine that we've already put into the atmosphere to head up into the stratosphere and do its worst to the ozone that should be there, and the stratospheric ozone layer will, we hope, be able to recover. In 2016, the size of the ozone hole was a bit smaller than the 1991 to 2016 average. That's a bit of a smaller area, and a bit less deep than in 2015, so moving in the right direction but there's still much less ozone up there than the pre-CFCs average. Early days.

Jonathan Shanklin, one of the team who carried out the first groundbreaking measurements of the ozone hole, turned out to be no general dogsbody after all. Shanklin recently commented that the most important lesson of the ozone hole is not that we can fix it by making some changes to what we put in aerosol cans, important as this is. There is a bigger picture, which is that we humans can make major changes to our thin and fragile atmosphere pretty quickly. Once we've done that, it takes a long time to sort out the problems that we cause. The ozone hole was a relatively straightforward problem to identify and solve – although we're still waiting for changes made over the past thirty years to make an appreciable difference to stratospheric ozone: as I said, early days. Dealing with climate change is a lot harder: we know there's a problem, but unlike the relatively clean fix for the ozone hole, there's no readily available solution. We continue to push carbon dioxide and other greenhouse gases into the atmosphere, with no credible strategy in place for finding alternatives.

OZONE IN THE TROPOSPHERE

Coming closer to ground level, we find ozone being formed in a different way. The ingredients for forming ozone down near the ground are oxides of nitrogen and volatile organic compounds – widely known as VOCs. Add a load of sunshine and perhaps a dash of particulate matter, and you have the perfect ingredients to form photochemical smog. These photochemical episodes are very different in nature to the murky European wintertime smogs that we'll investigate in Chapter 6, and first became apparent in parts of the world where there were lots of road vehicles with fabulous chrome exhaust pipes, and plenty of sunshine: welcome to 1950s California.

Smog in Los Angeles[42]

In the late 1940s, residents of Southern California started reporting eye irritation, and farmers found that their crops were being damaged in ways that they could not account for. Enter Arie Jan Haagen-Smit, a Dutch flavour chemist who had recently been appointed a professor of bio-organic chemistry at California Institute of Technology.[43] Haagen-Smit was the right man in the right place at the right time: his laboratory was in Pasadena, just downwind of Los Angeles; traffic and associated pollution was growing dramatically; and he brought experience of European smogs and a fresh perspective on a growing problem in environmental science.

As a man with a nose for a smell, Haagen-Smit realised that the Los Angeles smogs did not smell like the European smogs. The Californian smogs had a characteristic bleach-like odour, very different to the acrid smell of sulphur-driven winter smogs. This led Haagen-Smit to investigate the contaminants in the local Los Angeles smog by collecting them in the same apparatus that he was experienced in using to collect and analyse the flavour components of pineapple. The contaminants which turned up in the cold traps were partially oxidised organic chemicals. So it looked like the process causing the air pollution problems was the progressive oxidation of hydrocarbons in some way. Haagen-Smit investigated the effects of these chemicals on crops, and found similar symptoms to those reported by farmers. Perhaps there was a bit of good fortune involved, because the partially oxidised hydrocarbons were produced by reacting gasoline with the most readily available oxidant in the laboratory, which just happened to be ozone. Whether by accident or design, this laboratory method reproduced what was going on in the atmosphere very well.

Before too long, the reaction mechanisms responsible for producing ozone, and partially oxidised hydrocarbons, and crop damage, eye irritation and a murky haze were being unravelled. The starting point is the action of sunshine on nitrogen dioxide, and any ozone that might already be present:

$$NO_2 + \text{sunlight} \leftrightarrow NO + O$$
$$O_3 + \text{sunlight} \leftrightarrow O_2 + O$$

These are reversible reactions, so quite often the oxygen atoms will re-form ozone and nitrogen dioxide. The net effect is to convert a bit of light energy to a bit of heat energy. However, the energetic oxygen atoms formed in this process can also react with water molecules to form reactive hydroxyl radicals:

$$H_2O + O \rightarrow OH + OH$$

The hydroxyl radical is a pretty simple guy: one oxygen atom and one hydrogen atom, giving it the chemical formula OH. It's a water molecule (H_2O) with one of the hydrogens removed, or equivalently, an oxygen atom which has gained a hydrogen atom. In the lower atmosphere, OH radicals are formed from the reaction of particularly energetic oxygen atoms with water, as in the above reaction, or by the decomposition of organic peroxides.

Like the oxygen atom, the hydroxyl radical is very keen to react with the first likely molecule it comes across. This may be a carbon monoxide molecule, or an organic molecule, and it kicks off the process which forms ozone in the atmosphere down near ground level. VOCs are chemicals made up mainly of chains of carbon and hydrogen atoms, with some additional chemical groups which determine the reactivity of each chemical. For our purposes, we often denote a non-specific hydrocarbon (that is, any old hydrocarbon molecule) as "R–H" because the key reaction step involves removal of a hydrogen atom by the OH radical, to form water and a reactive hydrocarbon radical, R:

$$R\text{–}H + OH \rightarrow R + H_2O$$

The hydrocarbon radical quickly forms a peroxy radical by reacting with an oxygen molecule – well, there's a lot of oxygen around for it to react with:

$$R + O_2 \rightarrow RO_2$$

The main route for the organic peroxy radical is to react with nitric oxide to form nitrogen dioxide:

$$RO_2 + NO \rightarrow RO + NO_2$$

Nitrogen dioxide is then available to react with sunlight and restart the process. The fate of the RO radical is more complex and variable, but typically the net result is the formation of an HO_2

radical. This is likely to end up as another OH radical, and another molecule of nitrogen dioxide, again restarting the process:

$$RO + O_2 \rightarrow R'CHO + HO_2$$
$$HO_2 + NO \rightarrow OH + NO_2$$

As long as there are hydrocarbons, oxides of nitrogen and sunlight present, the overall effect is a progressive oxidation of hydrocarbons. The process generates more and more ozone as the radicals cycle round from OH to R to RO_2 to RO to HO_2 and back to OH. Every time this happens, NO is oxidised to NO_2 which pretty quickly results in another molecule of ozone.

And that's the problem – the reactions cycle round and round, creating ozone as a by-product on every loop. The result is rapid formation of ozone when the conditions are right: plenty of hydrocarbons, plenty of nitrogen dioxide, and some sunshine – the unholy trinity of the messy haze which we've come to know as photochemical smog. This runaway increase in ozone levels is why summertime smogs cause eye irritation and damage to crops. Doing something about it, however, is difficult. One of the best ways of reducing exposure to ozone is to live somewhere that isn't very sunny, but try telling a Californian not to live in California. Apart from that rather drastic measure, the strategies for control focus on reducing emissions of hydrocarbons and oxides of nitrogen, the two main ingredients for brewing up an airshed full of ozone.

When the mechanisms for forming ozone were understood and action was needed to do something about it, once again, Arie Jan Haagen-Smit was the right man at the right place. As the first chairman of the California Air Resources Board, he led the first initiatives to reduce vehicle emissions. Emissions limits on hydrocarbons started being applied in the mid-1960s, ten years after the chemical mechanisms were understood. This might seem a long time between identifying a problem and taking action, but any experience of dealing with the environmental impacts of commercial activity will tell you that ten years is actually pretty speedy. For example, evidence on the health effects of smoking emerged during the first half of the twentieth century, and was accepted even by industry insiders by the mid 1950s.[44] But the relatively modest step of requiring a health warning to be placed on cigarette packets wasn't taken in the UK until 1971. And cigarettes weren't banned from sale until – oh, hang on, they're still available, despite the well-known health effects. So a ten year timetable for action to reduce NO_x and hydrocarbon emissions from cars in California after the evidence became available isn't bad going.

From the 1970s onwards, emissions from road vehicles slowly started to reduce. The process accelerated with the introduction of the three-way catalytic convertor, which has done us proud in reducing the contribution of petrol cars to both these substances. We have now reached the point where they've given us everything they are going to, and we will need to find new ways of reducing emissions to make further progress on low-level ozone.

THAT OZONE SMELL

The bottom line with low-level ozone is that it's quite a reactive chemical, which oxidises many of the substances that it might come in contact with. Substances like crops, for example, or specialist plants in natural ecosystems, or your throat and lungs. Make no mistake, it's nasty stuff. You can sometimes smell ozone in the air near to an office photocopier or printer. That's because the bright lights or high voltage discharges used in a photocopier can enable the first reaction to take place – the splitting of a di-oxygen (O_2) molecule into two oxygen atoms.

The same smell crops up if you're brave enough to sniff the air just after a lightning strike. Not the aroma of charred golfer, welcome though this might be, but the smell of ozone formed by passing an electrical spark through the atmosphere. A German chemist, Christian Schönbein was the first to isolate this gas in 1840, and he named it "ozone", which means simply "smell". The name "ozone" with all those "o"s doesn't have anything to do with oxygen, as I had rashly assumed: it actually comes from the Greek word "ozein" which simply means "to smell". Although the basic make-up of the atmosphere comprising nitrogen and oxygen was well known by this time, the chemical formula of ozone was not established until 1865. Perhaps because of its distinctive smell, or maybe because of its antiseptic action as a strong oxidiser, ozone was widely thought to be beneficial to health from its discovery onwards.

To me, the odour of ozone is not a pleasant one. Yet if we go back to the nineteenth century, ozone would have been firmly in the public consciousness, not only as an antiseptic, but as a generally Good Thing for health and well-being. People would go to the seaside, or even to the Ozone Bathing Pool in Croydon, to seek out ozone for its health-giving qualities. The characteristic tang of the seaside (at least around the UK) was attributed to ozone. People thought that it was the smell of "smell". It sounded clean, scientific and restorative. Ozone was the stuff that would do you a power of good, and help you to recover from all sorts of nineteenth century diseases, which are, thank goodness, less popular nowadays. In Lyme Regis, for example, a street called Ozone Terrace was built down by the sea in the 1890s, and it's still there today. Not for the last time, town planners may not have got it quite right when naming their latest street after a reactive, poisonous gas with an unpleasant smell. Maybe if Christian Schönbein had followed the approach to naming chemicals that his compatriots took with nitrogen, ozone might have ended up being called "giftigstoff" or "poisonous material." And Lyme Regis Town Council might have been slightly less keen on naming its new promenade Giftigstoff Terrace.

But as you may well know already, the characteristic seaside smell is not in fact ozone: the seaside is not usually characterised by the refreshing aroma of office reproduction equipment. The seaside smell is mainly caused by organic sulphides from rotting seaweed. It's easy to understand why the zingy, fresh-sounding "ozone" was a better marketing ploy in the competitive world of early twentieth century seaside resorts, than highlighting rotting seaweed or smelly sulphides. Nowadays, of course, in this information-rich society, with our well-established and widespread understanding of the atmospheric chemistry and health effects of ozone, no one really believes that ozone is good for your health. But wait! Here is an item in *The Independent* newspaper in 2008 about the delights of the Isle of Man: *"The island is re-modelling itself for the modern breed of active, ozone-seeking visitors who bring more to their holiday than a bucket and spade."*[45] And in *The Sunday Times* from 2016: *"Close your eyes. Can you smell a hint of ozone? Feel the stirrings of a sea breeze in your hair? Are you experiencing an unaccountable desire for a donkey ride? In interiors, the seaside season has started."*[46] *The Sunday Times* article is headlined *"Just add sunshine, says Katrina Burroughs"*. Yes indeed, just add sunshine to trigger the chemical processes which result in a hint of ozone and a dash of photochemical smog, with the exciting opportunity to experience a seasonal soupçon of asthma. And right now, I could book myself a holiday in the delightful looking Ozone Retreat, on Ozone Street, Victor Harbor, South Australia. I can only hope it's as good as it looks, and doesn't live up to its name. Perhaps ironically, Ozone Retreat is about as close a holiday destination as you can get to the Antarctic ozone hole, where ozone has indeed been in retreat for well over thirty years.

DEALING WITH OZONE

So I think we've now set the record straight. Ozone, definitely not a pleasant chemical, and to be avoided where possible. But let's not forget that, although it's bad for us up close, ozone in the stratosphere does a great job of absorbing most of the sun's damaging ultraviolet light. This has enabled us to evolve to live in a world where hardly any of the most damaging sunlight in the wavelengths between 100 and 300 nanometres reaches us. That's great as long as that's the world we live in. We need stratospheric ozone to carry on soaking up UV radiation for us, so that we can carry on going outside in the sun in charming places like Lyme Regis, the Isle of Man, Victor Harbor, or even Croydon. And we came perilously close to losing the precious layer of ozone.

So ideally, we should do our best to leave plenty of ozone up there in the stratosphere, while keeping levels of ozone down here in the troposphere to the minimum. What a shame that we've ended up doing the exact reverse: resulting in a hole in the ozone layer up there in the stratosphere, and at the same time giving us high levels of ozone throughout the world down here at ground level. Just what we didn't want!

The problem with ground-level ozone is that there are hardly any sources of ozone itself which contribute to levels in the atmosphere. So we can't just put limits on ozone emissions, like we can with some other pollutants. This doesn't mean that human activity makes no contribution to ozone levels – far from it. But ground-level ozone is an indirect result of emissions into the atmosphere, followed by chemical reactions in the atmosphere.

The levels of ozone at ground level are determined by the amount of oxides of nitrogen and VOCs in the atmosphere, and by the intensity of sunlight, which I might in a moment of weakness have referred to as an "unholy trinity". During the day, there is a continuous cycling between ozone, nitric oxide and nitrogen dioxide, driven by the action of sunlight on nitrogen dioxide which leads to the re-formation of ozone. At night, of course, there is no sunlight, so once the sun goes down, ozone is on a one-way street, when it reacts with as much nitric oxide as it can find.

$$O_3 + NO \rightarrow O_2 + NO_2$$

Without any sunlight, there's no reverse process resulting in the formation of ozone. So, if there's more nitric oxide than ozone in the atmosphere, then the ozone all disappears at night. If there isn't enough nitric oxide, then there might still be some ozone around after dark. So be careful what you breathe.

These are complex processes – the formation of ground-level ozone depends on all kinds of factors including the intensity of sunlight, the presence of natural substances and man-made pollutants in the atmosphere, and rates of mixing and transportation. And that's before we get on to the significance of reactions taking place on the surfaces of water droplets, or indeed experimental equipment, just to make life tougher for poor hard-working atmospheric scientists. Predicting ozone levels is like predicting the weather, but with several additional layers of complexity which makes it a lot harder.

One important consequence of these complex atmospheric processes is that they take time. This means that the highest ozone levels often don't occur where the highest levels of precursors are present. Emissions of NO_x and VOCs are often greatest in urban areas, with road traffic typically making a major contribution. The rapid reaction of nitric oxide (NO) with ozone means that ozone levels are actually often pretty low in urban areas. As the emissions are carried away by the wind, particularly under conditions of strong sunlight, the reactions which result in ozone formation take place, and result in high levels of ozone downwind of the source areas. For example, the areas experiencing the highest ozone levels in the UK are East Anglia, Wales and the south coast. Conversely, areas such as central London and the cities of Liverpool, Manchester, Bradford and Leeds get off comparatively lightly.[47]

Fortunately, we can now use computer models to represent these processes at an overall level with reasonable confidence. Other factors being equal (which of course they never area, but this does get us started), levels of ozone depend strongly on the levels of both oxides of nitrogen, and VOCs. This is illustrated in in the next diagram.

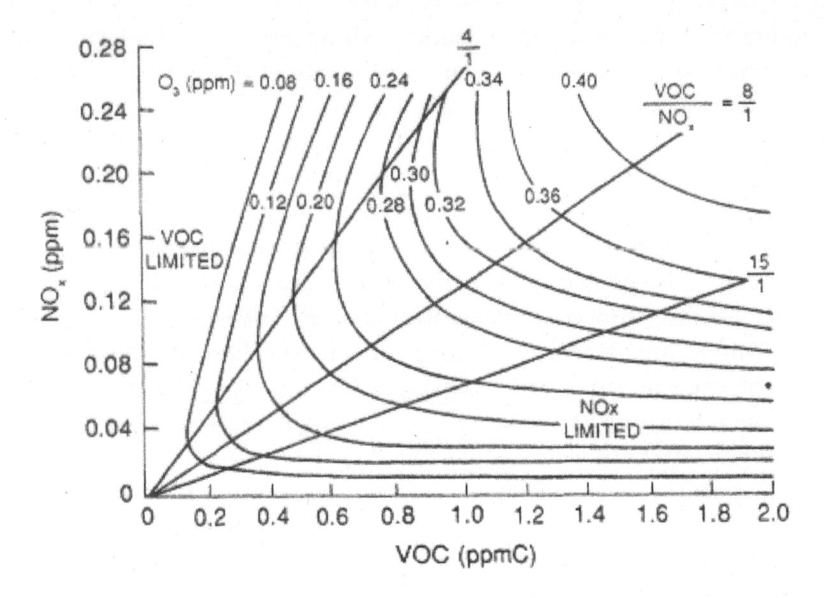

Interaction between ozone, NOx and volatile organic compounds[48]

The diagram shows the typical relationship between ozone levels and levels of NO_x and volatile organic compounds. If the concentration of volatile organic compounds is about eight times the concentration of NO_x (that's the line running diagonally down the middle of the graph), then you can reduce levels of ozone by reducing concentrations of either NO_x, or volatile organic compounds, or both.

However, if you have much more volatile organic compounds present compared to NO_x, then increasing or decreasing the levels of VOCs doesn't make much difference to levels of ozone – this is the region marked "NO_x limited" in the diagram. Under these circumstances, the way to secure a quick benefit in ozone levels is to reduce NO_x levels.

Similarly, if you have a lot more NO_x present compared to VOCs, then reducing NO_x levels won't have much benefit in reducing ozone. This would be typical of an urban environment.

In this situation, to improve ozone levels, the most effective approach would be to reduce VOC levels.

That's what the science tells us. However, as we are now looking at what can be done to improve air quality, we have to keep an eye on the practicalities. The science tells us that in an area where NO_x levels are relatively low (for example, a rural area), we should look to reduce NO_x further to make a difference to ozone levels. Any reduction in NO_x levels would deliver a corresponding benefit in reduced ozone levels. The problem is that it is always tough (by which I essentially mean expensive) to reduce emissions of a pollutant which is already at low levels. There are not likely to be many options available for further reductions in emissions. OK, so we could look at reducing VOC levels, which might be a more attractive option, but the problem here is that you are likely to need a substantial reduction in emissions before you see any benefit in ozone levels.

Conversely, in an area where NO_x levels are high, there could be cost-effective opportunities to deliver reductions in NO_x emissions. For example, encouraging people to reduce the use of private cars, introducing a charge on high-emitting vehicles, requiring developers to provide electric vehicle charging points, and encouraging the use of zero-emissions technologies to provide space and water heating. That's all great, and if it's possible to secure a sufficient reduction in emissions, this would eventually provide an improvement in ozone levels. However, in the earlier stages of an emissions reduction programme in an area of high NO_x levels, there are unlikely to be benefits in ozone levels, and in fact, ozone levels might even increase due to the reduced availability of nitric oxide to remove ozone from the atmosphere.

So there are no easy answers for dealing with ground-level ozone. It's a public relations challenge before we even start, because of its dual role in protecting and poisoning us. The United States Environmental Protection Agency came up with the catchy jingle "Good up high, bad nearby" (please feel free to add your own tune) to help people understand and remember what was going on with ozone. Not ideal, I'd have thought, because you could easily get it back to front: "Bad up high, good nearby...", no that's no good. My best suggestion is "In the stratosphere, keeps your skin clear; terrestrialise, hurts your eyes". I know, if anything that's worse that the EPA's effort, so please do suggest a snappier alternative. Anyway, this dual role means that we need one lot of policies to restore ozone high up in the atmosphere, and another lot of policies with a very different focus, to reduce ozone at ground level. It's fair to say that dealing with the ozone layer has had a higher public profile than dealing with ground-level ozone, to the extent that there's even an International Day for the Preservation of the Ozone Layer on 16 September every year. In 2016, the strapline was: "Ozone and climate: Restored by a world united". I think mine's better – at least it rhymes, and it's a bit of a stretch to suggest that we have successfully dealt with both ozone depletion and climate change. But there's no International Day for the Elimination of Tropospheric Ozone. To cut our exposure to ozone, we need to control emissions of different pollutants, usually from cities which are not themselves strongly affected by ozone, and which may not even be in the same country. And if we do manage to cut emissions of the pollutants which result in ozone formation, the diagram above shows that it's possible that it won't make much difference to ozone levels until we can deliver really major improvements.

These difficulties go some way to explaining why there has been a lack of progress in addressing the health effects of ground-level ozone. The Institute for Health Metrics and Evaluation and the Health Effects Institute estimate that exposure to ozone is responsible for about a quarter of a million deaths per year worldwide. This puts ozone at number thirty-three in the list of global disease factors, with a similar impact on health to occupational injury. What is equally disturbing is that the estimated burden of disease associated with ozone has increased steadily from about 150,000 per year in 1990 to 250,000 per year in 2015, and shows no sign of slowing down.[49] About two-thirds of this increase occurred in India, reflecting the combination of increased emissions, strong sunlight to drive the photochemical processes, and high population densities.

Anything that causes a quarter of a million deaths per year, and rising, should be a matter of concern. As we take steps to deal with urban air quality with its increasing focus on fine particulate matter, it's important that ozone – and particularly its impact on rural communities worldwide unfortunate enough to be located downwind of sources of pollution – isn't forgotten.

CHAPTER 5

WHATEVER HAPPENED TO ACID RAIN?

Within A Few Hundred Kilometres

AIR POLLUTION AND NATURE CONSERVATION

Back in the introduction, I mentioned that the motivation for my PhD project was to understand a bit more about the effect of naturally-occurring organic sulphur compounds on deforestation in Central and Northern Europe. And how it turned out that there wasn't any such effect. Indeed, the Norwegian Environment Ministry says unequivocally that: "Acid rain is mainly caused by combustion of fossil fuels."[50] In Norway, damage to forests was pretty minimal: devastation of forests by acid rain was much more widespread in Central Europe, probably linked to the extensive, and intensive, combustion of coal in Western and Central Europe during the middle of the twentieth century. These forests are now generally recovering from the worst effects of acid rain, as combustion of coal in Europe continues to decline. Between 1990 and 2014, for example, Europe-wide emissions of sulphur dioxide, one of the key contributors to acid rain, decreased by almost 90%.[51]

So let's continue our journey through the atmosphere – we've despatched the global climate, and taken a look at the regional scale problems with too much, and too little ozone. Coming closer to home, within a distance of a few hundred kilometres, brings us to acid rain. Acid rain has had a longer-lasting effect on the quality of rivers and lakes in Norway, and indeed throughout Europe, than it has on the forests. Acidic water bodies are still widespread throughout the southern half of Norway. Acid deposition continues to have an important impact on sensitive habitat sites throughout the UK, although the effects are now more subtle than the widespread deforestation of a few decades ago. Nowadays, the problem is less to do with sulphur dioxide from power stations, and more to do with our old foe, oxides of nitrogen. As we all now know, oxides of nitrogen come, not just from power stations, commercial and domestic combustion, but also from road traffic. These sources of emissions are generally reducing (which is good news), but only slowly, and not nearly as fast as sulphur dioxide. Oxides of nitrogen emissions from Europe decreased by just over half between 1990 and 2016, compared to the 90% reduction in sulphur dioxide.

Little did I know, when working on my PhD between 1989 and 1992, that the impacts of air pollutants on natural ecosystems would stay with me for the next twenty-five years. And here's me with no more than a limited grasp on the difference between a weed and a valuable flowering plant in our garden. The slow pace of decline of NO_x emissions means that deposition of acids, both as acid rain and by direct transfer from the atmosphere to vegetation, remains a major problem. If anything, dealing with the impacts of air pollution on ecosystems has grown in importance, as our knowledge and understanding of these impacts has developed. In fact, we in the UK take our natural habitats so seriously that protecting nature conservation sites often puts more of a constraint on new developments than protecting people. The big deals for valuable ecosystems are not so much to do with air pollution in the atmosphere, although that can be important – it's more to do with what the pollution does after it hits the ground. And the big deals are nitrogen deposition and acid deposition.

TOO MUCH NITROGEN, TOO MUCH ACID

So, what are the problems? Nitrogen deposition is slightly easier to get to grips with, at least as far as the maths goes. The starting point here is biodiversity: the amazing multiplicity of ways that biological systems have evolved to cope with and exploit different environments throughout the world. All plants need nitrogen to thrive – that's why we put fertilisers into the soil to encourage garden flowers and crops to grow. But some parts of the natural world are very low in nitrogen, and here some plants have evolved with specific strategies to obtain the nitrogen they need. So, for example, the Venus flytrap in the southern United States, and the worldwide family of drosera, or sundews, have adapted to gain their nitrogen and other nutrients by catching and digesting insects. Delicious.

Venus flytrap (originally from North and South Carolina, but this one in Cambridge University Botanic Garden)[52]

Plants like the sundew have evolved to live in low-nitrogen niche environments around the world which other plants can't manage, so you won't find them anywhere near a field which gets a regular dose of pig slurry. The trouble is that we humans have come along over the past hundred years or so, and started burning quite a lot of fossil fuels. Even if we're burning these fuels in ovens, boilers, cars and power stations a long way away from the nitrogen-deficient areas, the atmosphere is a good way of transporting the oxides of nitrogen from where they are emitted to these often remote habitats. Then it may rain on the sundew plant (it tends to rain quite a lot in, to pick a random place where you might come across a sundew, the Scottish moors) and a bit of the nitrogen dioxide dissolves in the rain, and lands on the low-nitrogen environment. So there's an increase in available nitrogen, and that means that any old plant can come along and live there. Before you know where you are, you've got a bit more grass and no more sundews – a process known as "eutrophication".

Acid deposition is similar in some ways – it arises from chemicals including sulphur dioxide, nitrogen dioxide, ammonia and hydrogen chloride in the atmosphere. These chemicals can be deposited directly onto vegetation, and they also dissolve in rainwater and land on plants and soils. The result is an increase in acidity and damage to ecosystems – particularly lichens and mosses, and woodlands.

Fortunately, we have now recognised these problems, and we're taking steps to deal with them. Damage caused by acid deposition has been recognised for over a hundred and fifty years, and the substantial reductions in sulphur dioxide emissions have helped to limit acid damage in recent years. However, emissions of oxides of nitrogen and ammonia continue to result in levels of acid and nitrogen deposition high enough to have long-term effects on sensitive habitat sites.

As well as these indirect effects of air pollution, habitat sites are also vulnerable to the direct effects of some air pollutants. The most critical of these is often ammonia. Ammonia is particularly important for habitat sites firstly because many plants can have their growth affected by this chemical – and in particular, some mosses and lichens. Lichens can be particularly sensitive to air pollution, because they don't have any roots, and so obtain all their nutrients directly from the air. This makes some lichen species highly susceptible to changes in the chemical composition of the atmosphere. The UK's safe level of airborne ammonia to protect sensitive lichens is one hundred and eighty times lower than the safe level for protection of human health. Wow, there are some sensitive lichens out there. And the second reason why ammonia is such a big deal is that it is released from agricultural activities and waste processes, which often take place in the countryside not very far away from protected habitat sites. Under these circumstances, any increase in ammonia levels could be a problem.

DEALING WITH IMPACTS ON BIODIVERSITY

Dealing with the direct and indirect effects of air pollutants on habitat sites is mainly about reducing source contributions. Whatever steps we take to reduce emissions from road traffic and other sources to improve air quality for people living near to and downwind of major sources of emissions will also improve air quality at sensitive habitat sites, even if these are further afield. Improving the environment for sensitive ecosystems and habitat sites isn't likely to be a major driver for widespread air quality improvements, but might be a fringe benefit from health and compliance-driven improvements in air quality. One problem is that the effects of air pollutants on sensitive habitats are subtle. An ecologist can't very easily go to a habitat site, and highlight damage or improvements caused by small changes and trends in air pollution. So any verification or assessment of air pollution impacts and benefits has to be indirect.

A secondary part of dealing with this problem is to do with new development. Local planning authorities and environmental regulators such as the Environment Agency in England now have to consider impacts on specifically designated habitat sites as part of their decision-making processes. The methods and criteria that we are required to use when carrying out these assessments can be extremely draconian. That's good, because it means that we have a lot of protection in place to stop individual developments making air pollution worse at some of our most valuable habitats in the UK. But of course, there has to be a balance, and the balance between environmental protection and allowing beneficial development swings substantially towards environmental protection for some habitat sites.

One example of this is a waste to energy plant at St Denis in Cornwall.

The Cornwall Energy Recovery Centre[53]

This is a middle-sized waste to energy plant, which is designed to process about 240,000 tonnes of residual waste per year. But it doesn't have a middle-sized chimney: it has two 120 metre chimney stacks, among the tallest stacks of any such plant. That's a third taller than Big Ben. Why are the chimney stacks so high? It's not because of the risk of emissions to local people, but because of the site's location close to some highly protected habitat sites.

The stacks are so high that, even at the nearby protected site, which comes within a couple of hundred metres of the chimneys, the impact would be minuscule. The stacks were designed in this way, so that the risks of impacts could be screened out without the need for a more detailed evaluation. But really, it's a sledgehammer to crack a nut: the sledgehammer is a pair of 120 metre stacks which can be seen for miles, and the nut is the avoidance of any possible impacts due to air pollution emissions from the waste incinerator on the nearby Breney Common, Goss Moor and Tregoss Moor habitat sites. They are lovely, remote places, inhabited by species such as the fragile marsh fritillary butterfly. The problem is that this 120 metre sledgehammer doesn't do anything about other more urgent threats facing these habitat sites, such as emissions from the A30 which runs adjacent to the site, as well as invasive species, nutrient enrichment from animal food and dung, and the use of fertilisers and pesticides.[54] Don't get me wrong, I'm very happy to see the moors protected, but building a more expensive, more visible stack than is needed isn't the way to go about it.

One challenge with trying to manage the effect of air pollution at protected sites is that it is usually pretty hard to see evidence for actual damage due to air pollution at actual habitat sites. There are often other threats to habitat sites such as overgrazing, water pollution and the presence of visitors and dogs. With all this going on, the subtle background effects of air pollution, maybe resulting in slightly fewer individual members of specially adapted species for low nitrogen-poor environments, are hard to discern. As a result, we end up dealing with these impacts through the medium of arithmetic – one number to describe an impact, divided by another number to describe the sensitivity of the site. This can take away from the immediacy of the effects of air pollution on sensitive ecosystems.

A BIT OF FIELDWORK

But however hidden the impacts are, air pollution continues to have a real impact on biodiversity at many sensitive sites. We can tell this because we have an understanding from field and laboratory experiments of how air pollution affects different plants, and we have good information on the levels of air pollution at every habitat site in the country. In fact, we can even estimate air pollution levels using the presence of the most sensitive species as an indicator. One way is to record the presence of nitrogen-sensitive and nitrogen-tolerant lichen on tree trunks and branches. These observations can be turned into a score to evaluate levels of nitrogen in the atmosphere. As someone who loves the countryside, but hasn't studied biology since the age of fourteen, I nervously took the Field Studies Council's guide to assessing air quality using a lichen based index out to some woodland near my home, together with a compass, a magnifying glass and a tape measure, and had a go. Now, the first problem is that you have to be able to identify oak and beech trees, which was a bit of a challenge, but I think I got it right by the simple expedient of only doing oak trees. Even I know what an oak tree looks like. After that, I was able to make some tentative identifications of lichen on the trunks and branches of some pretty substantial oak trees, and use the guide to classify air quality in terms of the presence of reactive nitrogen in the atmosphere – which means, oxides of nitrogen and ammonia.

The tree trunks told me some good news: that air quality at that particular woodland is "clean" – plenty of nitrogen-sensitive species on the tree trunks, and not so many nitrogen-tolerant ones. But then when I looked at the branches, although there were fewer lichen to be seen covering much smaller areas, they seemed to suggest that the air quality is "polluted". The guide suggests that this means that air quality at the site is deteriorating, which I guess makes sense if there aren't so many nitrogen-sensitive species on the newer growth. I cross-checked this by taking a look at the air monitoring records for ammonia, available from the UK's air quality archive. And blow me down, at the nearest monitoring site (about thirty kilometres away), measured ammonia levels have been going up pretty steadily by about 2.3% per year since 1997. The measured annual average ammonia level is currently lurking somewhere below the safe level for most vegetation, and above the safe level for sensitive lichens. On the current trend, it'll tip over the higher guideline at this particular wood in about 2040, so we'd better get a shift on if we're going to sort things out. It looks like the lichen and the national network measurements are telling the same story, and it isn't a particularly happy story. In my corner of the country, ammonia levels are not too bad at present, but seem to be getting worse. Elsewhere, ammonia levels are high enough to have direct impacts on habitat sites already.

DEALING WITH AMMONIA

Ammonia is proving to be a particularly intractable pollutant to deal with. It has multiple impacts on habitat sites, contributing to both acidification and eutrophication, as well as being directly toxic to plants. At some monitoring stations, levels are pretty constant, while at others, including my local, ammonia levels seem to be drifting up. That's a bit odd, because at the same time as measuring ammonia in the atmosphere, we also make estimates of the amount of ammonia emitted into the atmosphere from the UK as a whole. Over the twenty years or so between 1997 and 2015, our best estimate is that there was a consistent decrease in ammonia emissions of about 1% a year, with this rather feeble decrease coming mainly from the agriculture sector. It doesn't look like this is being reflected in what we measure in the atmosphere, and measurements don't lie, so it looks like we might need to take another look at these estimates. It's important legally, because we need to make sure we comply with limits set in the Emissions Ceiling Directive, which derive from the United Nations Convention on Long-Range Transboundary Air Pollution. As the UK has signed this United Nations Convention, we'll need to meet our emissions reductions obligations, come what may in our relationship with Europe. But the bigger picture is that the legal obligations are there to deliver improvements in the environment for both people and natural ecosystems.

Why is ammonia such a tough one, though? This curious little chemical doesn't come from the sources that we've grown used to as the ones which cause us air pollution problems – traffic, power stations, industrial activity and so on. Instead, the UK's national inventory suggests that the only show in town for ammonia emissions is farming, which accounts for about four-fifths of all the ammonia released into the atmosphere. As farms tend to be located in the countryside, they are handily close to the valuable and sensitive habitat sites which might be affected by ammonia. I mentioned that it's a curious chemical – the formula is NH_3, so that's a nitrogen atom and three hydrogen atoms. Ammonia forms an alkaline solution in water, which caused me some consternation when I was trying to work out how it could contribute to acid rain. But it does, through biochemical processes which result in the oxidation of alkaline ammonia to acidic nitrates. At much higher concentrations than we normally encounter in the atmosphere, ammonia has a very pungent smell. This led to its use in smelling salts to revive fainting ladies in Agatha Christie detective stories – indeed, Christie's story *The Blue Geranium* turns on the effect of ammonia from Mrs Pritchard's smelling salts on red flowery wallpaper. Yes, wallpaper can act as litmus paper (thereby fulfilling the prediction of Agatha Christie's fortune-teller Zarida), as long as you can convincingly plaster litmus paper onto the wall.

Coming back to the real world, the ammonia which is released from agriculture into the air is split pretty equally between emissions from handling manure, and emissions from soils. So half from raising animals and half from growing crops. Emissions from this kind of activity are hard to quantify accurately, and it's harder still to make inroads into reducing emissions from such widespread sources. Air quality measurements, field experiments, and surveys of habitat sites tell us that we should really try to do something to reduce ammonia from farming. But there are almost two hundred thousand farms in the UK,[55] all of them different, and if I've learned one thing from the past twenty-five years of studying air pollution impacts on habitat sites, it's that farmers will do what they want to do, not what some nerd from an office tells them they ought to do. And as growing crops and raising livestock are, approximately, natural activities, there will always be some releases of ammonia into the atmosphere as nitrogen cycles around through animal feed, into manure which becomes fertiliser for crops, from where it's taken up into the crops and back into feeds. Having said that, there are ways of cutting down ammonia emissions from agriculture. The most clear-cut methods are to do with managing animal feeds, and the ways that manure and animal slurries are stored and applied to the land as fertiliser,[56] and there are plenty more opportunities besides. It's farmers who have delivered the estimated 1% reduction in ammonia emissions year on year since 1997 – so that's a 20% reduction in agricultural emissions of ammonia over that twenty year period. Good work everyone, but the trend in the last couple of years has been back up again, and even a 20% reduction is more of a marginal improvement than a fundamental change. Delivering more fundamental improvements in emissions of ammonia from agriculture will take a combination of incentives, financial and technical support, and regulatory back-ups to tackle a wide range of sources and make sure that the whole industry works to the same standards as the best of today's farmers.

Acid rain was the problem that got me on my way in the world of atmospheric science in the 1980s. Back then, the effects were all too obvious. Despite making dramatic improvements in emissions from industry and road vehicles since then, it seems that air pollution is continuing to have subtle but important effects on many of our most precious ecosystems. As we saw with the ozone layer, this underlines how fragile our world and its atmosphere is, and how easy it would be to accidentally lose what we value most.

CHAPTER 6

AIR POLLUTION AT THE CITY SCALE

Within A Few Tens of Kilometres

Air quality varies dramatically, not to say wildly, from place to place. The two most obvious factors which influence air quality are: where the pollution comes from, and where it goes. On the one hand, the location and features of pollution sources are the starting point for dealing with air pollution (for example – how much pollution per second? Are emissions constant or varying? How high up is the source?). And on the other hand, the weather conditions affect where emissions from a particular source go, driven in particular by the wind direction and speed. That's not the whole story, but it's a big part of it. So, for example, odours are likely to be strongest close to the source, whether that's the downstairs toilet or downwind of a field where a farmer is throwing pig slurry into the air. Where solid fuels are burnt on open fires and stoves in the home, you're likely to find high levels of airborne particulate matter containing some pretty nasty chemicals, both indoors and outdoors. Levels of air pollution from road traffic are typically highest in urban areas – particularly cities with congested streets and less than satisfactory controls on vehicle emissions. In a big city, you often have roads and other sources of pollution in every direction, resulting in high pollution levels whatever the wind direction.

URBAN AIR QUALITY MANAGEMENT IN THE 17TH CENTURY

Air pollution is a topic which has interested scientists for centuries, although maybe it didn't get quite the level of attention afforded to turning base metals into gold, or the use of leeches to treat diseases due to imbalance of the humours. People have been highly conscious of what they were breathing in for centuries, without necessarily knowing what it all was. In a pamphlet published in 1661, the writer John Evelyn considered air pollution in London. "*Fumifugium, or, The inconveniencie of the aer and smoak of London dissipated together with some remedies humbly proposed by J.E. esq. to His Sacred Majestie, and to the Parliament now assembled*" is one of the first analyses of the causes of air pollution and sets out proposals for delivering improvements in air quality.

Evelyn was perhaps unfortunate in trying to get to grips with the problems of atmospheric pollution at a time when the science of chemistry itself was just shaking off the shackles of alchemy. With its mystical and unsystematic approach to understanding the make-up of the world around us, alchemy had held back progress in chemistry for centuries. It's no surprise that physics, biology and mathematics developed and widened their insights into the world around us throughout the middle ages, while alchemists found it hard to make much progress in turning anything that wasn't gold into gold. By the time Evelyn turned his attention to air pollution, chemistry was in its infancy – indeed, Robert Boyle's formal distinction between chemistry and alchemy *The Sceptical Chymist* was published in 1661, the same year as *Fumifugium*.

Evelyn was working from simple visual and olfactory observation of industrial pollution – what he could see and smell. But Evelyn carried out his analysis a hundred years too soon to have any understanding of the chemical make-up of the atmosphere. He was able to identify the burning of sea coal as a

FUMIFUGIUM:

OR

The Inconveniencie of the AER

AND

SMOAK of LONDON

DISSIPATED.

TOGETHER

With ſome REMEDIES humbly

PROPOSED

By *J. E.* Eſq;

To His Sacred MAJESTIE,

AND

To the PARLIAMENT now Aſſembled.

Publiſhed by His Majeſties Command.

Lucret. l. 5.

Carbonúmque gravis vis, atque odor inſinuatur
Quam facile in cerebrum? ——

LONDON,
Printed by *W. Godbid* for *Gabriel Bedel*, and *Thomas Collins*, and are to be ſold at their Shop at the *Middle Temple* Gate neer *Temple-Bar*. *M. D C. L X I.*

John Evelyn's 1661 pamphlet *Fumifugium*[57]

major source of dust and soot in the atmosphere, and also drew attention to tallow and blood odours from chandlers and butchers. Evelyn's proposed solution was to move the noxious industries that burn sea coal and produce excessive odour eastwards down the river Thames. This would be downwind of the prevailing westerly wind direction, so that smoke and odours would have much less influence on air quality in the parts of London where all the important people lived. And it would also be a job-creation opportunity, because river boatmen would be needed to transport all the useful products from these smoky, smelly industries back to the city. He invited the new Restoration king, Charles II, to take these steps to clean the city, suggesting that in doing this, Charles would perhaps embody the role of Christ in clearing out traders from the temple in Jerusalem.[58]

17TH CENTURY SOLUTIONS IN THE 21ST CENTURY

If Evelyn's solution had been taken on board, how different would London's air pollution problems be? The heavy industry in central London would have been relocated to areas such as my old manor (well, it was my manor up to the age of nine) of Hornchurch in Essex and Dartford in Kent – and, in the end, that's roughly what happened, even if it did take several hundred years for the Ford Motor Company to move into Dagenham. The last major industrial process in central London was a combined heat and power plant operated by Citigen at Charterhouse Street. This closed in 2014 to be replaced with a much smaller and less polluting plant, and now there are no processes large enough to be regulated by the Environment Agency in central London, finally fulfilling John Evelyn's vision of an industry-free environment for the oligarchy.

Although there's no longer any large-scale industrial activity in central London, recent growth in the use of diesel-fired plants throughout London to generate electricity to meet peak demands does perhaps raise concerns that we've forgotten what we ever learnt from John Evelyn. This unexpected change in direction is a consequence of the trend in electricity generating capacity away from large power stations and towards small-scale and more intermittent generating capacity. This is generally a positive trend – in particular, renewables now account for about a quarter of all electricity generated in the UK. However, from time to time, the energy demand spikes – for example, during the adverts in Coronation Street, or at half time during an England football match, as fans watching on TV go off to drown their sorrows in a mug of tea. The renewable energy sources can't always be ramped up to give sufficient capacity to cover these spikes in demand, and indeed might be off-line when the demand peaks, so there is a market for reliable short-term energy suppliers. The need for these suppliers is particularly strong in cities, where the short-term demand is most intense. The attractive contracts on offer encourages suppliers to install and use cheap and reliable diesel-fired generators in city centres. This is great for meeting short-term demand, but you'd have to say, something of a step backwards for urban air quality.

What will make matters worse is that the electricity demand is only going to increase as the market share of electric vehicles continues to rise. It's great not to have air pollution from electric vehicles on the road, but that electricity has to come from somewhere and it's not all clean and green – not yet, anyway. Meanwhile, the residents of east London continue to have to put up with relatively poor air quality, due partly to ongoing industrial activity in the suburbs, but more to pollution from road traffic in their local area and throughout London.

URBAN AIR QUALITY MANAGEMENT IN THE 19TH AND 20TH CENTURIES

Let's go back a century or two. We first started to get to grips with understanding how air pollution affects our health in the middle of the twentieth century. One of the often forgotten features of mid-twentieth century life, along with vinyl records of pop songs name-checking the atmosphere, is the infamous smog which affected many industrialised areas of Europe. The portmanteau word "smog" evocatively describes the particularly unpleasant mixture of smoke and fog which used to result from widespread burning of coal when the weather conditions were cold, calm, damp and generally miserable. Smogs had, of course, been a fact of life for many years, going back to John Evelyn's description of London smogs caused by widespread burning of sea coal. A couple of hundred years after that, things hadn't really improved: in *Bleak House*, Charles Dickens used the term "London Particular" to refer to a thick smog:

> *"Only round the corner," said Mr Guppy. "We just twist up Chancery-lane and cut along Holborn, and there we are in four minutes time, as near as a toucher. This is about a London particular* now, *ain't it, miss?" He seemed quite delighted with it on my account.*
>
> *"The fog is very dense indeed!" said I.*
>
> *"Not that it affects you, though, I am sure," said Mr Guppy, putting up the steps. "On the contrary, it seems to do you good, miss, judging from your appearance."*
>
> (Charles Dickens, *Bleak House*, Volume 1)

Arthur Conan Doyle describes thick, discoloured fogs in a number of Sherlock Holmes short stories around the turn of the twentieth century – although the term "pea-souper", along with "elementary my dear Watson", was never used by Conan Doyle. The Impressionist artist Claude Monet painted the Houses of Parliament in the fog during his visit to London in 1904, capturing the effects of sunlight and fog on the view from his lodgings in a series of dramatic and sometimes (to my untutored eye) murky paintings.

These cold weather smogs were originally characterised by particles and sulphur dioxide from burning coal in businesses and industrial processes. By the nineteenth century, the London population was large enough that domestic coal burning was a major contributor to smogs, alongside the industrial sources described by Evelyn. During the winter, a lot of coal was burnt particularly on domestic fires, and when this coincided with still weather conditions, the result was a pretty unpleasant combination of smoke, sulphur dioxide and fog.

SMOGS OF THE 1950s

In the UK, a series of smogs in the early 1950s was enough to force action to be taken to deal with these winter episodes. My sister-in-law recalls the smog turning her beautiful white fur hat a dirty grey colour while trying to find the way home through smoggy Leicester back in the 1950s (white fur hats: another forgotten feature of twentieth century Britain). The obviously unpleasant nature of the smogs was part of the reason for action, but in fact public attention after the smog of December 1952 was focused much more on the risk of crime during a smog, the vexatious cancellation of sporting fixtures (at White City greyhound races, the dog literally couldn't see the rabbit), and the death of some prize cattle at the Smithfield show at Earls Court.[59] If the smog was responsible for ill-health and premature mortality in cattle, it seems surprising that there weren't questions around whether similar effects might happen among the human population. Perhaps because Twitter was still sixty years away, there wasn't a public outcry, at least not at first.

Before long, however, the General Register Office published an estimate that the smog had resulted in about four thousand premature deaths. This estimate was obtained by comparing the death rate during the smog of 5 – 9 December 1952 with the death rate before and after this time, and the death rate during the same period of 1951.[60] The figure of four thousand deaths was probably

The London smogs of the early 1950s[57]

an underestimate, not least because it only considered the period of the smog itself, and not the period immediately after the smog when the death rate remained elevated above the normal rate for several days. This one episode may actually have resulted in twelve thousand deaths.[61] London smogs could no longer be ignored, or treated as an endearing quirk of London life, with complexion-enhancing properties.

Epidemiology (the study of the distribution and causes of diseases) was coming of age at this time, with influential studies

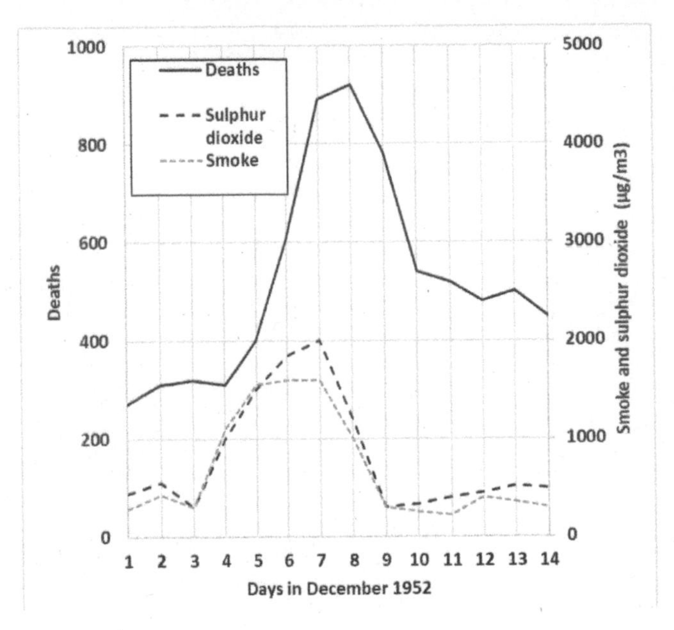

Pollution and deaths in London during the December 1952 smog[62]

of the effect of smoking on health published by Richard Doll and Austin Bradford Hill in 1950 and 1954. In between these studies, the groundbreaking epidemiological research which followed the 1952 episode culminated in the 1956 Clean Air Act. This act has been updated several times, but its key provisions remain in place: restrictions on the use of solid fuels in specified urban areas; a ban on the release of black smoke from commercial premises; and (from 1968) minimum height requirements for industrial chimneys. It took a while for these measures to take effect – another episode in 1962 was responsible for about 750 premature deaths – but finally the controls introduced by the Clean Air Acts started to reduce and eventually eliminate the occurrence of severe winter smog in the UK. So much so that the old smoke and sulphur dioxide network station in central Manchester recorded levels of sulphur dioxide in 2005 which were just 2.4% of the levels measured in 1962.

THE EFFECTS OF AIR POLLUTION ON HEALTH

Alongside the common sense and effective measures in the Clean Air Acts, wider changes in sources of air pollutants have been taking place. There has been a progressive reduction in industrial activity in urban centres and throughout the UK. This has been complemented by a more widespread move away from burning coal and oil in the manufacturing, power generation, commercial and domestic sectors, and towards the use of natural gas and renewable sources of energy. These fuels result in much less, or in some cases zero, sulphur dioxide and smoke. At the same time, the number of vehicle miles travelled increased fivefold over the thirty years between 1952 and 1984.[24]

The research around the London smogs showed that air pollution episodes can cause thousands of premature deaths, and we knew a lot about it half a century ago. But it turns out that those short-lived pollution episodes, dramatic as they can be, are not the biggest deal for the effects of air pollution on health. In the past few years, a lot of big numbers have been published which shed new and disturbing light on the effects of air pollution on health. From the BBC in 2016 alone, we had:[63]

"UK air pollution 'linked to 40,000 early deaths a year'"
"Air pollution 'causes 467,000 premature deaths a year in Europe'"
"Polluted air causes 5.5 million deaths a year new research says"
"Polluted air affects 92% of global population, says WHO"

That's a lot of early deaths. But there is no disease called "air-pollution-itis", so what are all these people dying of? These estimates are based on a statistical analysis of death rates, so these analyses don't themselves necessarily give us any insight into the actual effects of the air pollutants on our health. They just tell us that there are higher death rates in populations exposed to higher levels of pollution. We do know that the majority of the deaths from air pollution arise from fine particulate matter, $PM_{2.5}$, which gives us a clue as to how the pollution affects our health. Nitrogen dioxide and, to a lesser extent, ozone also contribute to premature deaths caused by air pollution.

Separate research then begins to explain how these pollutants affect our health. As we saw in Chapter 3, the very definition of $PM_{2.5}$ specifies that they are the respirable particles which can get deep into the lungs. The presence of these particles causes irritation and inflammation to the lining of the lungs. For most people, this would be painful or inconvenient, perhaps by restricting the efficiency of the lungs and making it a bit harder to breathe. Unpleasant, but not life-threatening for most of us. But for some people, who have a pre-existing disease, or who are elderly or otherwise vulnerable, this additional strain on the respiratory and cardiac system could be much more serious. It's this group of people who may perhaps not survive an illness that they would otherwise have managed, resulting in a small increase in the risk of death across a large population. You can't be sure whether any individual death has been hastened by air pollution, although you might have a strong suspicion, but you can pick it up as an overall increase in the mortality rate when populations experience higher levels of air pollution compared to other groups of people.

HOW BIG IS THE PROBLEM?

How big is the problem? As the BBC report focusing on premature deaths in Europe tersely says, "it's pretty bad".[63] The risk can't be identified or specified for individuals, but it is possible to estimate the increased risk at a population level. The UK Committee on the Medical Effects of Air Pollutants (COMEAP) estimated in 2010 that all-cause mortality would increase by a factor of between 1.01 and 1.12 per 10 $\mu g/m^3$ change in the annual average airborne $PM_{2.5}$ concentration.[64] COMEAP's best estimate was a factor of 1.06 per 10 $\mu g/m^3$ change in $PM_{2.5}$ levels – that is, a 6% increase in all-cause mortality per 10 $\mu g/m^3$ change.

That's a lot of numbers, and it takes a bit of work to understand what they mean. The big picture is that it stacks up to millions of early deaths worldwide each year. In the UK, exposure to $PM_{2.5}$ is responsible for almost 5% of all deaths, and worldwide it's over 10%. At a local scale, the differences from place to place are pretty small. That's because you need a change of 10 $\mu g/m^3$ in annual average $PM_{2.5}$ levels to produce an increase of 6% in the mortality rate, and this is actually quite a large change in concentration. The highest and the lowest values recorded in 2016 at the UK's seventy-three national network monitoring stations were 17 $\mu g/m^3$ by the side of the A450 in Birmingham, and 3 $\mu g/m^3$ at Auchencorth Moss in the Scottish Borders. That's a difference of just 14 $\mu g/m^3$ between the highest and lowest values recorded anywhere in the UK. Putting it another way, the differences between $PM_{2.5}$ levels from one place to another are pretty small, so the differences in mortality due to exposure to $PM_{2.5}$ from place to place would also be pretty small. Nevertheless, adding up the effects of everyone's exposure to $PM_{2.5}$ worldwide makes exposure to air pollution the single largest environmental cause of premature death in the world, resulting in tens of thousands of premature deaths in the UK, and millions worldwide. Elsewhere in the world, particularly in large urban areas, much higher levels of $PM_{2.5}$ are recorded, and that's why the effects on mortality are even higher.

The COMEAP publication, and its 2009 predecessor, was something of a watershed moment. COMEAP showed that we really needed to be looking seriously at $PM_{2.5}$, and not just focusing on complying with air quality standards for nitrogen dioxide, PM_{10} and ozone. This was one of the triggers for a substantial expansion of $PM_{2.5}$ monitoring: up to 2008, the number of monitoring stations in the UK national network was in single figures. In 2008, this jumped to forty-four monitoring stations, and has continued to rise up to seventy-nine stations in 2018. Before we started taking $PM_{2.5}$ seriously, a lot of the focus of air quality policy and action had been on nitrogen dioxide and ozone, along with PM_{10} (there were around fifty national network stations measuring PM_{10} throughout the 2000s). Now we know, or think we know, that most of the health impact of air pollution is due to $PM_{2.5}$. Views on the contribution of nitrogen dioxide to health burdens continue to change as new research becomes available. The main difficulty is that levels of air pollutants often go up and down together, so when you carry out a study designed to measure the effect of air pollution on health, you often don't know which pollutant, or combination of pollutants, is causing the effect that you observe. However, there is now more or less a consensus that $PM_{2.5}$ is responsible for most of the mortality burden of air pollution, with smaller contributions from nitrogen dioxide and ozone.

We don't exactly know how those effects are shared out among the population. COMEAP estimated that exposure to man-made $PM_{2.5}$ in the UK results in a shortening of life for all of us of about 6 months on average. But this could reflect a bigger effect for a smaller number of people, while the rest of us are relatively unaffected. This uncertainty means that we should, strictly speaking, describe the effect on mortality as equivalent to about 30,000 early deaths per year in the UK: we can't really be confident how the effect on health is distributed through the population. It could be a lot of people with a slight shortening of life, or a smaller number of people who each experience a severe effect, or somewhere in between.

Heart disease and lung conditions are most commonly linked to air pollution, and have the most widespread effects on health. About 60% of the effect of air pollutants on mortality is due to ischaemic heart disease. Cardiovascular disease, lung cancer and chronic obstructive pulmonary disease also contribute to the observed increase in mortality.[65] Additionally, the liver, spleen, central nervous system, brain, and reproductive system can also be damaged by exposure to air pollution.

ACUTE HEALTH EFFECTS

We are probably more familiar with the acute health effects of air pollution, such as you might experience during an air pollution episode – bearing in mind that "acute" means "short lived", and not necessarily "severe". Although short-lived health effects can, of course, be very severe. The acute health effects of air pollution are many and varied. Among the most severe is the risk of triggering asthma attacks by high levels of the irritants ozone and nitrogen dioxide. This can happen during episodes of very high levels of these pollutants. Less scary, but still unpleasant, those same photochemical pollutants can make your eyes itchy or sore, or make your nose run, or make you feel wheezy. All pretty nasty effects of air pollution. Air pollution episodes are an occasional fact of life in rainy Northern Europe, and much more common in parts of the world where emissions from cities, industries or agricultural burning are cooked up with a dose of sunlight. We could also include smells in this category – a usually intermittent, short-lived problem, but as well as the experience of the odour itself, smells can have knock-on effects. An unwelcome smell such as decomposing food or poorly treated sewage might make you feel nauseous and also cause secondary problems, such as headaches, stress or sleeplessness.

And then there's hay fever, or allergic rhinitis as I should say. I'm not sure that we should strictly speaking include pollen as an air pollutant, but pollen is certainly strongly influenced by human activity – planting and encouraging grass to grow, for example. I'm a hay fever sufferer myself, but for me at least, it's an inconvenience rather than a serious health problem. The very first day that I experienced hay fever symptoms was when I was sitting an O level history exam aged sixteen. Fortunately, I passed the exam, but it wasn't a very pleasant experience for me or for my fellow students with surnames beginning with B who were unlucky enough to sit either side of me, particularly because, being sixteen years old, I'd gone into the exam without a hanky. And it still catches me out every year when I head out into the countryside in early June, and remember, while trying to scratch the inside of my eyeballs, that I should have been taking the medication for a few days. The incidence of hay fever in the UK increased dramatically over the 1970s, 80s and 90s,[66] round about the time I was doing O levels. Since then, the trend has levelled off, leaving us with a remarkably high incidence of between 10% and 30% of adults in the UK suffering from hay fever.[67]

Our understanding of the acute effects on health of air pollution has become well established over the past decades, but the much more substantial and insidious burden of long-term exposure to air pollution has emerged and gained widespread acceptance only in the past few years.

AIR POLLUTION FROM NEW ROADS

When I started my first proper job in 1992, we in the UK were just coming to the end of an era of building new roads. That was just twenty years after the first environmental consultancies had been established in the UK. And air pollution was not the first thing on the environmental industry agenda: there was a lot of groundwork to be done on contaminated land, water resources and waste management. During the 1980s and early 1990s, air quality started to figure as a small part of the environmental assessment work carried out by the fledgling environmental consultancy industry in relation to new road construction. Detailed environmental studies, including air quality analysis, were needed before a new road could go ahead, and the Department of Transport laid down methodologies for how these studies should be carried out. These methods seemed to ensure that environmental assessments were huge and detailed documents which invariably and mysteriously demonstrated that the new road would be great, and any impacts insignificant by comparison.

Perhaps there was just a smidgeon of truth in the *Yes Minister* description of the UK's road building policies. Civil Servant Bernard Woolley explains to his Minister Jim Hacker why there are two decent motorways which would take you from London to Oxford. The answer was that all departmental Permanent Secretaries studied at Oxford, and like to enjoy a splendid college dinner – hence the motorways. So why were the M4 and the M40 to Oxford completed so many years before the M11 which would take you to Cambridge (not to mention ports and other more practical destinations)? Simple: the Department of Transport never had a Permanent Secretary from Cambridge.

Whether or not it's true that the costs and environmental impacts of a new motorway are outweighed by a decent college dinner, and you can't help but feel it might be, two of the last new road construction schemes in the UK (the M40 extension from Oxford to Birmingham, and the A1–M1 link road, now part of the A14) had just been opened the year before I started work. I worked for a consultancy which was part of a large engineering firm, and the engineers were struggling to come to terms with the likelihood that we would not be building many more exciting new cuttings through hills like the rich grassland of Twyford Down near Winchester. Their lives in future would be dedicated to less dramatic tasks, like designing motorway maintenance and improvement projects while maintaining a two carriageway flow in each direction. Even so, most of my work for the first few years was focused on highways improvements, but with a definite shift away from the construction of new roads and towards improving the standards of existing infrastructure. Some of the schemes I worked on have been very beneficial – in particular, improvements to what's now the M60 motorway around Manchester – but in a few other cases, I was very glad that the schemes never saw the light of day.

As part of the Manchester orbital motorway improvements, pretty much my first job was to run an air quality monitoring survey in south Manchester. I had a rude introduction to the stresses of fieldwork when I turned up to one of our monitoring stations one morning. The station comprised two Horiba instruments conveniently installed on a shelf in a boathouse at the Sale Water Park.

A Horiba NO_x, SOx and oxygen monitor circa 1992[68]

I arrived at the site to carry out the weekly calibration checks, to find a large hole in the wall and the police investigating a ram raid carried out the previous night. The thieves had got in through the wall underneath the shelf containing our instruments, and successfully made off with a couple of outboard motors. They had presumably taken a look at the two instruments buzzing away above their heads, and decided to quit while they were ahead, because the monitors were still running fine. I guess outboard motors had a higher retail value in the pubs of south Manchester than air monitoring instruments. While it was a testament to the robustness of the Horiba kit, unfortunately, all I could offer the police was an indication of environmental levels of nitrogen dioxide and carbon monoxide during the raid.

GETTING TO GRIPS WITH URBAN AIR QUALITY

Alongside the decline in road transportation projects in the early 1990s came fundamental changes to the regulation of environmental pollution in the UK. The Environmental Protection Act of 1990 was a key turning point in environmental regulation, introducing far-reaching controls on contaminated land and waste management – and also regulation of industrial processes. This required operators of larger industries to apply for an operating licence which considered the potential impacts of emissions to all environmental media: air, land and water – hence the name "integrated pollution control" for this permitting system. So it was a good time to be an air quality specialist with an interest in evaluating emissions from industrial processes. I enjoyed unique opportunities to accompany our technician and ex-mining engineer Louis Barlow up a range of chimneys and vents throughout the UK. There is a unique form of misery which results from the combination of performing technical tasks requiring manual dexterity and strict conformance to written protocols while standing on a small metal platform at the top of a ladder in 50 mile an hour sleet. I particularly remember a happy day spent at the Tioxide paint factory on Teesside, trying not to lose a small white filter paper covered in white particles in the white snow, while my glasses were alternately blurred with white clouds of steam and more white snow.

And there was one infamous night on the way back home to Huddersfield from a day of monitoring at Whitehaven in Cumbria. 25 January 1995. How can I be so sure of the date? It was the same night that Eric Cantona, having been wound up all evening by Crystal Palace fans, took a flying kung-fu kick at one of them on the way down the tunnel after being sent off. It would have been a mildly interesting event I suppose, in the absence of any other news that evening, although not one that warrants a twentieth anniversary retrospective show on BBC Radio Five Live, you might have thought. But there was some other news that evening: along with several hundred other travellers, I was in a long, stationary queue of traffic heading up the M62 from Manchester towards Yorkshire. The snow had started as we travelled back through Lancashire, and traffic on the M62 had got slower and slower before finally grinding to a halt halfway up the Pennines. I was pretty sure I would be OK, because the radio reports were saying that the M62 was still open, and this was, after all, the motorway that they said would never be closed by the weather. But no, a not very large but very badly timed snowfall closed the motorway heading up out of Manchester, and I and many other motorists were well and truly stuck. Of course, I kept listening to the radio to hear updates of how the emergency services would be keeping us alive during this winter night on the side of a bare mountain – snowploughs? Food parcels? Helicopter rescue? But no. Every radio station was giving blanket coverage to a kick from a Frenchman to a bloke from south London.

In the end, we didn't all die: we eventually turned round and went the wrong way down the motorway back to the services, where I was able to get hold of a plate of baked beans and black pudding (Which I had to pay for! Even though I nearly died!) and doze for the rest of the night before heading home the next morning. Maybe you had to be there to appreciate the true severity of the conditions.

So much for the joys of industrial air pollution monitoring. Even more significant for air quality in the UK was the publication by the government of a little-noticed document in 1995, entitled *Air Quality: Meeting the Challenge*. We already had air quality standards and guidelines in the UK, but for the first time, this document set out some systematic, strategic policies which were aimed at making sure that controls on air pollution were sufficient to enable us to comply with the standards. For a document produced by a Conservative government, it's always seemed to me to be a, dare I say laudably, non-Conservative approach. It proposed an expanded role for local authorities in assessing air quality, identifying areas where air pollutants might be above the relevant standards, and taking action to deliver compliance. Yes, an increased role for local authorities, and a new administrative process, designed to achieve environmental targets. While it was a bit short on commitments for specific action, nothing like this had featured in the Conservative's 1992 Manifesto (yes, honestly, I've read it), and it was the start of two decades of air quality management in the UK. I'm not enough of a political insider to know how this much-needed and ground-breaking strategic approach to dealing with air quality saw the light of day, but I guess we have a serious pollution episode in December 1991, and John Gummer, now Lord Deben, to thank for its introduction. The 1991 episode lasted for four days, with nitrogen dioxide levels up to 800 $\mu g/m^3$, the highest recorded by the automatic monitoring network before or since,[69] and the first time that air pollution became front page news since the smogs of the 1950s and 1960s.

Not long after this consultation document had been published, the Environment Act 1995 became law. Yes, yet another fundamental change to environmental regulation in the UK in the 1990s. Heady times. As well as setting up the Environment Agency, Scottish Environmental Protection Agency and National Park authorities, the Environment Act required the development of a National Air Quality Strategy, and set up the system of air quality management envisaged in *Air Quality: Meeting the Challenge*. This embodied and enabled a move from a piecemeal approach to air quality towards a managed process aimed at delivering air quality standards throughout the country. No longer was air quality assessment limited to assessing individual road schemes and industrial sites. No, local authorities were required to measure and assess air quality in their areas, and keep checking it, and consult with their neighbours, and do something about it if they found a problem. As a quality management approach, you couldn't really fault it. In 1990, there were thirteen air quality monitoring stations in the UK national network measuring levels of nitrogen dioxide. By 2000, thanks to the introduction of local air quality management, this had increased to eighty-seven monitoring stations.

By the time air quality management got up and running properly, the European Union had helpfully stepped in to set some more demanding and legally binding air quality standards, and lay down minimum requirements for air quality monitoring. Long may these requirements continue. So by the early 2000s, we were no longer thinking about air quality just in terms of Twyford Down and the Newbury bypass. The processes of local air quality management were getting under way to measure, assess and deal with air quality problems throughout the country.

PROGRESS REPORT ON AIR QUALITY MANAGEMENT

It would be nice to report that all the hard work over twenty years carried out to identify air quality problems, design solutions, consult with local communities and implement the required measures have very satisfactorily dealt with air pollution so that there are no longer any problems. Sadly, that isn't quite the case. While there have been successes in dealing with air pollution hotspots, air quality standards aren't yet achieved in thirty-seven of the UK's forty-three reporting zones,[70] which is not a great result from a twenty year programme. On the plus side, things are already looking good in six zones (Scottish Borders, Highland, Blackpool, Preston, Brighton and Northern Ireland), and all we have to do is wait until 2027. By then, the measures that we've already got in place will deliver compliance with the standards in all forty-three zones, except for central London. It's not just a matter of sitting back and waiting, of course: the duty imposed by the European directives is to achieve the air quality standards in the shortest possible time. More on this in Chapter 10.

It seems to me that there are two reasons why urban air quality continues to be a problem, to the extent that the UK and other countries are facing legal proceedings in relation to their failure to achieve air quality standards quickly enough. Firstly, the process of air quality management had the required muscle in relation to mandatory air quality standards (tick), measuring and assessing air quality (tick) and methods for requiring industrial sources of air pollution to clean up their act, where needed (tick). But there was no similarly effective means of dealing with traffic pollution, particularly where this traffic was using motorways and trunk roads. While some measures have been taken, such as the use of variable speed limits to reduce congestion on motorways, these are driven by the roads authorities' objectives of improving safety and journey times, not by any requirement to deliver improvements in air quality. It's handy when these objectives coincide, but air quality concerns alone have been spectacularly unsuccessful in securing action from the highways authorities. Secondly, measures to reduce the impact of road pollution often involve restrictions or additional costs for motorists – or both. In a democratic society, every motorist is a voter – and most voters are motorists. Very few motorists like to see restrictions placed on when, where and what they can drive. Particularly when these sacrifices are required for something as intangible as an air quality improvement. So, while public perceptions are now changing, it has taken a lot of bottle for any elected body to take effective measures to improve the air quality impact of road traffic.

The biggest exception to this is probably London, with its Low Emission Zone covering most of the area inside the M25 – some 300 square kilometres, the largest LEZ in the world – and the Congestion Charge zone in the centre. It's difficult to pinpoint what benefit these measures have had on air quality – not because they haven't had an effect, but just because it's hard to assess what would have happened without them. The current Mayor of London, Sadiq Khan, has made air quality one of the priority areas for his administration, and has announced plans for further charges on diesel vehicles, and the implementation of an Ultra Low Emission Zone. Let's hope it'll work: central London currently experiences the highest pollution levels in the country. In 2017, the air quality standard for annual mean nitrogen dioxide concentrations was exceeded at almost half of London's air quality monitoring stations, and the highest levels experienced by Londoners were almost two and a half times the air quality standard, so there is still a long way to go.[71]

Let's have a look at one success story and one ongoing challenge for urban air quality.

A SUCCESS STORY: LEADED FUEL

Who remembers four star petrol? That is, motor fuel with lead in it. Lead additives were put into petrol for many years to enable engines to operate at higher compression ratios, thereby improving control of the combustion process, eliminating knock and improving power output and efficiency. The problem is that lead is a toxic substance with a wide range of health impacts, including harm to the developing nervous system in children resulting in attention deficit disorder, aggressive behaviour, and reduced intelligence. It is also a carcinogen. Burning automotive fuels which contain lead additives results in the release of lead in a form which is easily inhaled, at about the nose height of young children, often close to areas where children live. It's a terrible idea. There was perhaps a clue to the toxic nature of tetra-ethyl lead at an early stage: the guy who discovered the benefits of lead additives, Thomas Midgley, demonstrated the safety of tetra-ethyl lead at a press conference in 1924 by pouring it over his hands and inhaling its fresh aroma for a full minute. Even at that early stage, Midgley had already suffered from lead poisoning, and the occupational hazards of tetra-ethyl lead were so well known that it was known as "loony gas" among the workforce at the manufacturing plant.[72] But the bigger issue than occupational exposure of the workforce was the *"slow, unrelenting low-level exposure that was sure to occur as the use of leaded gasoline spread"*. Indeed it was: who knows how many children were affected by growing up in an atmosphere rich with lead from car exhaust pipes between the 1930s and 1990s, myself included. As Bill Bryson points out in his inspirational book *A Short History of Nearly Everything*, Midgley went on to be a member of the team which developed the first chlorofluorocarbon (CFC) refrigerant – a chemical which we now know to be implicated in both global warming and damage to the ozone layer. What an encore.

Back in the 1990s, I carried out an air quality study for a company called Associated Octel (now known as Innospec). Associated Octel was the world's leading manufacturer of tetraethyl lead, although by that time, lead was no longer used in most European countries. The issue we were looking at was that removing lead from fuel has to go hand in hand with the use of catalytic convertors. The lead in exhaust gases from cars using leaded fuels poisons the catalyst and stops it working, so you have to use lead-free fuel in cars with catalytic convertors. At the same time, in order to maintain the fuel octane rating, unleaded fuel has to have higher levels of aromatic chemicals like benzene and toluene in it. This results in higher emissions of benzene from the vehicle exhaust, and benzene is well known to be a potent human carcinogen. So you really need to have catalytic convertors fitted to vehicles which are using unleaded fuel. As a result, introducing catalytic convertors has had the additional benefit of hastening the end of leaded fuels wherever they are used.

The study we carried out for Associated Octel was to look at the effect of introducing unleaded fuel in countries where the vehicle fleet didn't have catalytic convertors fitted. The benefits of reducing lead in the atmosphere were well understood, but what would be the effect on benzene levels in a city like Bangkok where (at that time) they didn't have catalytic convertors? Well, not surprisingly, we found that introducing unleaded fuel without introducing catalytic convertors would result in high benzene levels. We weren't looking to take a view on whether the neural system damage due to lead exposure would be better or worse than the increased cancer rates due to benzene exposure: the study was designed to make sure that the transition away from leaded fuels was properly co-ordinated with the introduction of catalytic convertors. That's not a bad thing to do.

Nowadays, leaded fuel has been phased out worldwide. As of 2016, the United Nations Environment Programme thought that only three countries were still routinely using leaded fuels – Algeria, Yemen and Iraq.[73] It seems likely that the only country where use of leaded fuels might continue at any significant level is Yemen. The process of eliminating lead from vehicle fuels, since we first became aware of the environmental issues in the 1920s, has been painfully slow but it is finally pretty much complete. What finally brought about the removal of lead from fuel was not the harm it was doing to children's cognitive development, or the other health consequences of exposure to lead, but the need to eliminate lead from fuel used by cars with catalytic convertors. Once every car on the road had a catalytic convertor, the lead problem was more or less dealt with.

At least we can see the benefit of removing lead from vehicle fuels in measured levels of airborne lead in the UK. Leaded petrol was finally removed from sale in 1998, by which time the evidence of lead monitoring is that essentially no one was using it.

Levels of lead measured beside the M56 motorway in south Manchester were already falling by the mid-1980s, and the levels recorded since 2010 are well below 1% of the measured concentrations in 1984 and 1985. More importantly, the measured levels now comply with the European and World Health Organization guidelines by a considerable margin. Good news.

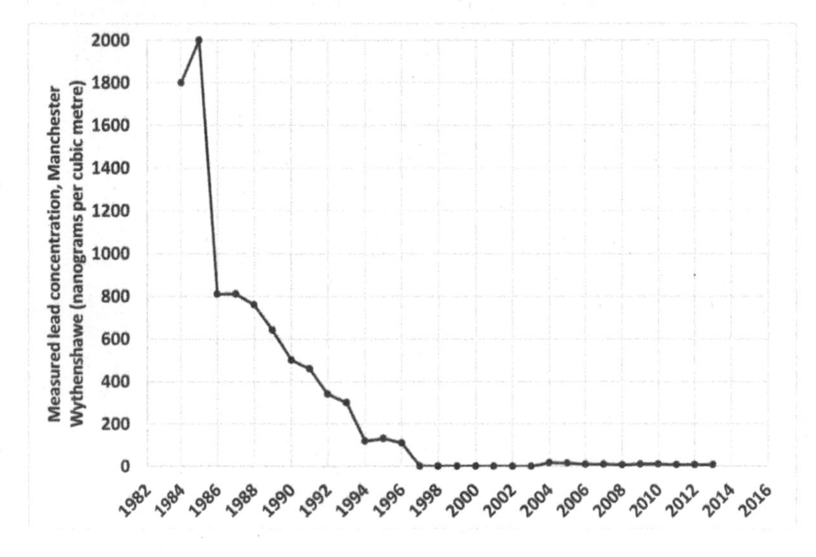

Measured airborne lead concentrations, Manchester Wythenshawe[34]

AN ONGOING CHALLENGE: $PM_{2.5}$

So, if we've dealt with the ozone hole, and got rid of leaded fuels, what's left in terms of air quality problems? The pollutants which continue to cause ill-health and premature deaths are fine particulate matter ($PM_{2.5}$), nitrogen dioxide and ozone. And the worst of these is the particulate matter. Whenever we see headlines like this from *The Independent* in 2017: "*UK air pollution deadlier than across half of western Europe, reveals WHO report*," the figures usually refer to the long-term health burden of exposure to airborne $PM_{2.5}$. Sometimes the figures also include nitrogen and ozone, with an effect on mortality around 10% of the effect of $PM_{2.5}$.

Let's have a look at how levels of $PM_{2.5}$ have evolved over the past few years in the UK. Because the evidence on the health effects of $PM_{2.5}$ has emerged relatively recently, we don't have a long record of measuring $PM_{2.5}$. The national air quality monitoring network only provides data back to 2009:

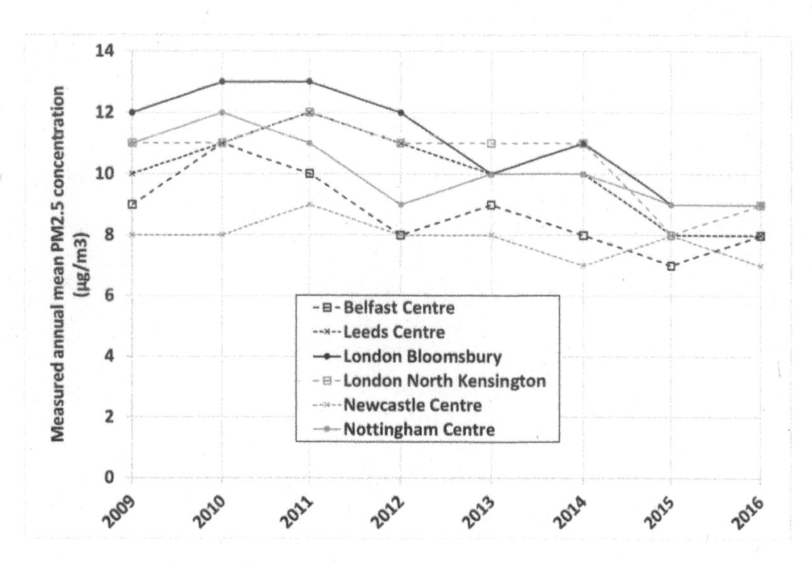

Trends in measured $PM_{2.5}$ levels in the UK[34]

It's hard to see anything of a trend in the $PM_{2.5}$ data for the past nine years – maybe a very slight decline, but certainly nothing you could be confident about. Fortunately, we have been measuring PM_{10} levels for a lot longer – here are the measured PM_{10} levels at the same monitoring stations.

The next figure shows measured levels of PM_{10} in urban centres of the UK have roughly halved over the period 1992 to 2016, with the decrease levelling off in recent years. It's likely that $PM_{2.5}$ levels have followed a similar trend over this longer period, but we don't have the data to be sure.

$PM_{2.5}$ is a bit of a messy pollutant. It's a lot easier to make sweeping generalisations about some of the other pollutants:

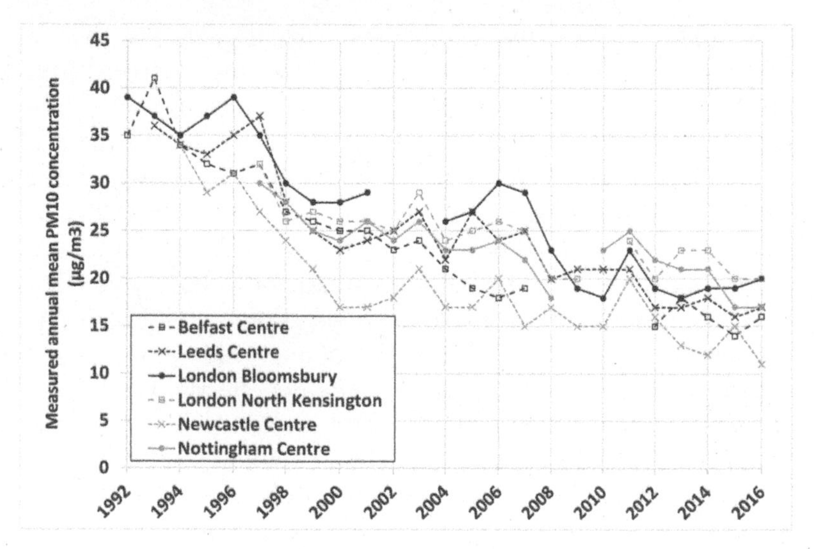

Trends in measured PM_{10} levels in the UK[34]

sulphur dioxide mostly comes from burning solid and liquid fuels, for example. In contrast, particulate matter has a much more disparate range of sources and life is made much more complicated by processes which form particulate matter in the atmosphere. Perhaps surprisingly, the biggest single source category for $PM_{2.5}$ in the national atmospheric emissions inventory for 2014 was – well, have a guess. I had a go, and I was completely wrong. Accounting for more than a quarter of emissions, it was domestic wood burning. Yes, the little wood burner in my sitting room, along with everyone else's, was by some distance the largest single source of particulate matter.

After that, the next highest source contributions were the use of fuel oil in shipping and agricultural straw burning (both 6% of total emissions) with road transport, my guess for the top source, trailing in at 5% of total $PM_{2.5}$. The top ten sources account for just over half of $PM_{2.5}$ emissions, with the balance made up of almost four hundred different sources. Clearly, dealing with $PM_{2.5}$ is not going to be easy: there aren't many easy wins which will deal with a hefty chunk of $PM_{2.5}$ emissions. Nevertheless, we have made some progress in more or less halving levels of fine particulate matter since 1992, as the graph shows. This is pretty consistent with a reduction in estimated $PM_{2.5}$ emissions from the UK as a whole, from 211 thousand tonnes in 1992 to 130 thousand tonnes in 2014. The biggest reductions in emissions of $PM_{2.5}$ over that period are from burning of coal in power stations and homes. At the same time, there has been a similar increase in emissions of $PM_{2.5}$ from domestic wood burning. The net effect of these three changes is a reduction of twenty-one thousand tonnes per year of $PM_{2.5}$ released to the atmosphere. The remaining sixty thousand tonnes of emissions reductions has been achieved through lots of smaller scale improvements. These include reduced emissions from industrial coal and oil burning, off-road machinery, foundries and diesel vehicles. The drivers for these changes were the wider-scale moves away from the use of coal and oil in industry, and of course the long-term declines in industrial activity in the UK. These improvements have also been beneficial in reducing the formation of particulate matter in the atmosphere.

On top of this, despite the best efforts of at least one car manufacturer, emissions from diesel vehicles have reduced over the past twenty-five years in response to international progress in setting limits on vehicle exhaust emissions. We now know that real-world emissions from all vehicles – particularly those produced by Volkswagen between 2009 and 2015 – are higher than the emissions tests and specifications might suggest. This divergence occurs for both legal and illegal reasons. Some vehicles produced by Volkswagen have been deliberately and illegally programmed to operate emissions controls only when being tested. Volkswagen uses a device called a Lean Nitrogen Oxides Trap to comply with emission limits on diesel vehicles. This system operates in two modes – removing NO_x from exhaust gases in a lean mode, then switching to a less efficient fuel-rich mode which allows the NO_x collected in the trap and the other exhaust gases to be converted to nitrogen, water and carbon dioxide. Because the fuel-rich mode is less efficient, Volkswagen disabled this mode for their vehicles on the road. Once the trap is full up with NO_x, it can't collect any more, so NO_x produced by the engine ended up being discharged from the exhaust pipe without any removal. Nightmare. Volkswagen developed a so-called "defeat device" which detected when a car was being tested, and would revert to periodically using the fuel-rich mode to clear the NO_x trap so that it would pass the tests when it had to.[74]

That was a specific, and definitely illegal, issue for Volkswagen diesel vehicles which came to light and was stopped in 2015. More generally, emissions from large numbers of vehicles (not just Volkswagens) have been found to be higher under real-world driving conditions than under test conditions. This doesn't come as a major surprise to air quality scientists, who have been trying to understand the discrepancies between claimed emissions performance and measured environmental levels of air pollutants for many years. And it's not necessarily illegal: vehicle manufacturers invariably design their engine systems to deliver compliance with the specific circumstances and requirements of emissions tests. They don't, and indeed can't, design their cars to minimise emissions under any other set of conditions which you might encounter on the school run, for example.

This means that vehicles have been emitting higher levels of air pollutants than the test data suggested they would for years, both legally (it's not necessarily illegal for real-world emissions to exceed test bed emissions) and illegally (you're definitely not allowed to fit a specific device to make sure that this happens), with all the environmental and health consequences of these higher emissions. The cost of the increased pollution in the US and Europe due to the defeat devices fitted by Volkswagen was estimated to be at least forty billion dollars.[75]

Notwithstanding a little difficulty with real-world vehicle emissions, at present, there are no strong drivers for action to further reduce levels of $PM_{2.5}$ in the UK. The air quality standard for $PM_{2.5}$ in England, Wales and Northern Ireland is 25 $\mu g/m^3$. The monitoring data for $PM_{2.5}$ shows that this standard is achieved pretty comfortably even in city centre locations. In Scotland, a more demanding standard of 10 $\mu g/m^3$ applies. Even this more demanding standard has been achieved at all monitoring stations in Scotland, except for a kerbside monitoring site in central Glasgow which stopped operating at the end of 2014. The kerbside site was located literally on the edge of Hope Street, next to Glasgow Central railway station. It's fair to say that this monitoring station was not representative of general public exposure to $PM_{2.5}$, or in fact to any individual exposure.

So there's no strong incentive to take additional steps to reduce concentrations further to comply with air quality standards in the UK – we're already there. The UK has also adopted an exposure reduction target for $PM_{2.5}$, with the objective of reducing urban background levels by 15% between 2010 and 2020. Monitoring data reported to the national networks suggests that we are on track to achieve this target too – in 2010, the average measured level at urban background sites was 10.0 $\mu g/m^3$, and by 2016, this had reduced to 7.8 $\mu g/m^3$, a reduction of 22%. This reduction is, of course, good news, but the cloud that goes with this silver lining is that, once again, there isn't any incentive to make further reductions in $PM_{2.5}$ levels. So levels of $PM_{2.5}$ here are looking pretty good, right? Right, except that they remain high enough to cause around 30,000 early deaths each year.

$PM_{2.5}$ FROM TRAFFIC

In their report on these early deaths, with the literally inspirational title of *Every Breath You Take*, the Royal College of Physicians and Royal College of Paediatrics and Child Health make fourteen recommendations for action to deal with this health burden of exposure to outdoor air pollution, and fine particulate matter in particular. These recommendations are useful suggestions for changing the way that people think and act. In terms of specific measures to reduce exposure to these pollutants, the report suggests promoting alternatives to petrol and diesel fuelled cars, putting the onus on polluters to reduce emissions, and leading by example in the National Health Service. The report also highlights actions that individuals can take, including reducing car use and improving the efficiency of domestic appliances.

Adding up all the different categories of road vehicles, petrol vehicles accounted for 0.2% of UK emissions of $PM_{2.5}$ in 2014. If we're serious about reducing $PM_{2.5}$, we need to look elsewhere: there is little point in exhorting drivers of petrol-engined vehicles to stop using their cars. Diesel vehicles, by contrast, accounted for 5% of $PM_{2.5}$ emissions. So there is a good case for moving away from diesel-fuelled transportation. Half the diesel vehicle $PM_{2.5}$ emissions were from cars, which is certainly an avoidable source of emissions in the medium term, as alternative technologies which don't emit as much pollution are available. So the first step should really be to make diesel vehicles unattractive to consumers. Over the vehicle replacement cycle of five to ten years, removing diesel cars from use could deliver a 2.4% reduction in $PM_{2.5}$ levels in the UK – maybe more in urban areas where traffic makes a greater contribution to air pollution. Two and a bit per cent doesn't sound a whole lot, but in the fragmented world of harmful $PM_{2.5}$, it's a good start.

$PM_{2.5}$ FROM DOMESTIC WOOD BURNING

We also need to look at domestic wood combustion, which comes hard to me as I've very much enjoyed having a small wood burning stove for the last seven or eight years. At more than a quarter of all the UK's $PM_{2.5}$ emissions in 2014, this is the one source that could really make a difference to $PM_{2.5}$ levels. Except that it's not one source, it's millions of individual sources. Estimating emissions from domestic wood combustion is fraught with difficulty. There is some data out there which enables us to calculate how much particulate matter is released when wood is burnt in domestic burners, but it's a bit of a stab in the dark to estimate how well people operate their fires. If you get it wrong, by burning unseasoned wood for example, the combustion process can be very inefficient and result in increases not just in $PM_{2.5}$, but also in emissions of some pretty nasty chemicals. That's one reason for the recent *Ready to Burn* campaign in the UK, encouraging recreational arsonists like me to use properly seasoned solid fuels.

Even when a household wood burner is operated correctly, a quick peer and a sniff at a domestic chimney will show visible and smellable smoke particles, presumably accompanied by an unhealthy dose of $PM_{2.5}$ particles which are too small to be seen. And in most cases, these burners are a leisure accessory rather than a necessity. Most of the houses with wood burners are already fully equipped with central heating, my own included. One way to make a real dent in those figures for premature deaths due to air pollution would be to ban, or restrict, or control the use of wood burners in domestic properties, particularly those located in urban areas where there are plenty of people nearby to be affected by whatever comes out of the chimney. The UK's latest Clean Air Strategy does indeed pay attention to the use of stoves and open fires in the home. While it stops short of a ban on domestic wood burning, plans are afoot to limit the sale of fuels and stoves to "the cleanest" and "the very cleanest" respectively. However, if you've got a dirtyish stove in your home already, you'll be able to carry on using it to burn the cleanest fuels. I'm pretty sure more stringent restrictions would be unpopular with the recreational arson demographic, but they could maybe could maybe save several thousand thousand lives a year. Isn't that a price worth paying?

$PM_{2.5}$ FROM INDUSTRY

Industrial processes and power stations also contribute almost a quarter to $PM_{2.5}$ emissions from the UK. This is a very disparate sector, with emissions coming from both fuel combustion and manufacturing processes. The larger industrial plants are regulated under the Environmental Permitting regulations. This process sets benchmarks for achievable emissions, with site-specific limits set in a permit for each regulated site. Emission limits tend to get lower over time, so we can expect a slow ongoing reduction in emissions from industrial processes, but don't hold your breath, even to avoid breathing in $PM_{2.5}$: changes are likely to be marginal and slow to work through.

$PM_{2.5}$ FROM SHIPPING AND AIRCRAFT

A source that I've only mentioned briefly is international transport – shipping and aircraft. That's remiss of me, because shipping is forecast to be an increasingly important source of air pollution in the coming years. And that's not necessarily because emissions are going up, it's more because they are not going down. Most other sources of emissions to air are subject to progressively tightening controls, so the proportion of $PM_{2.5}$ coming from ships and planes is forecast to go up, even if the amounts stay more or less the same. There are exceptions, like the growth in $PM_{2.5}$ emissions from domestic wood burning, but emissions of most pollutants from most sources continue to decline. Except for international transport. The reasons for this lie in the difficulties of securing international agreements on emissions controls. In the context of controlling greenhouse gases, the international community has been seeking limits on carbon dioxide emissions from aircraft for many years. The problem is that, once an aircraft takes off, no one country has responsibility for, or control of, its emissions to the atmosphere. It's not in any country's interest to unilaterally place additional requirements on aircraft in relation to greenhouse gas emissions, even if the mechanisms were in place to deliver and regulate emissions controls. Unless there is concerted international action, any country acting on its own would only succeed in driving international aviation to other countries where the restrictions and costs are less demanding on the airline. So reducing emissions from international aviation comes down to voluntary international agreements. For air pollutants, that's perhaps less of an immediate concern because the pollution emitted by an aircraft once it has gained any height in the atmosphere is not likely to contribute significantly to the levels experienced at ground level. The same isn't true for greenhouse gas emissions, of course: the pollutants contribute to global warming whether they are emitted on the tarmac at Heathrow or high above the Atlantic Ocean on the way to JFK.

International shipping emissions are perhaps more of a concern for air quality. Ships tend to emit their pollutants pretty close to sea level. For much of their voyages, ships are some distance away from land, but they can nevertheless make a significant contribution to air pollution – for example, by adding to the mix of pollutants which generate ozone through photochemical processes, as well as the direct impact of emissions when they are in harbour or close to the shore. As we've seen, sulphur dioxide emissions from international shipping doubled between 1990 and 2011 so that this sector now accounts for one seventh of global emissions. International shipping accounts for almost a tenth of UK $PM_{2.5}$ emissions, and 2% of global PM_{10} emissions, roughly the same as road transport. A recent study found that just fifteen of the world's biggest ships emit as much sulphur dioxide as all the world's 760 million cars put together – an astonishing finding, although perhaps it's as much to do with the extremely low emissions of sulphur dioxide from cars these days as it is to do with high emissions from shipping.

Nevertheless, emissions from shipping sure are high. Although emission controls are applied to shipping in international waters, compared to the limits which apply to land-based sources, these controls are amazingly lax. The International Maritime Organization has set a limit of 3.5% sulphur in fuel oil used in ships. Three point five per cent! That's a lot: a hefty chunk of the fuel that ships are burning is pure sulphur. It's not surprising that they emit a lot of sulphur dioxide as a result. By way of contrast, this limit is thirty-five times higher than the limit of 0.1% sulphur in diesel fuel used on land throughout Europe. The same is true of other pollutants: much of the emissions from shipping would not be permitted to occur in inland waters, or from similar sized engines on land. So it's not like there are technological constraints which result in such generous emissions from shipping: no, there are readily available measures to enable ships to operate with lower emissions of most of the headline pollutants. It's a matter of fuel availability and the costs associated with operating to more demanding emissions limits. Change is afoot – the limit on the sulphur content of marine fuels will tighten in the next few years, and the limits on emissions of oxides of nitrogen from newly constructed ships are also more reasonable, though still way above the equivalent limits for land-based diesel engines. There aren't any specific limits on emissions of particulate matter from international shipping, other than the limits on the sulphur content of marine fuels. And the new limits on NO_x emissions will take a while to work through the fleet, with the average lifetime of a container ship somewhere around twenty-five or thirty years.

Remote monitoring techniques are opening up new ways to check up on emissions from individual ships, which will make enforcing tighter limits on shipping emissions more straightforward. There is much to be gained from tightening limits on shipping emissions: the European Environment Agency estimated in 2013 that international shipping contributes between 5% and 10% to airborne $PM_{2.5}$ levels across much of the UK. If we could eliminate this as a significant source, it could enable us to avoid another few thousand deaths each year. So if we have to pay a bit more to import zips from China or bananas from South America – again, maybe that's worth it.

$PM_{2.5}$ FROM AGRICULTURE

Next up comes agriculture, also contributing about 10% to $PM_{2.5}$ emissions from the UK. There are parallels here with the situation of domestic wood burning. More than half the agricultural contribution to $PM_{2.5}$ is from the combustion of straw for heat and electricity production. This is an activity which has more than doubled since 1992 to be the main contributor to agricultural $PM_{2.5}$ emissions. Straw is an attractive option as a cheap source of fuel in an agricultural area, but it is a difficult fuel to manage, and emissions of particulate matter are about forty times higher per unit of energy generated than emissions from combustion of fuel oil.[76] Perhaps there should be a shift of focus away from combusting straw in small-scale boilers, and towards the use of cleaner-burning fuels. Using straw as a fuel in agricultural boilers is definitely a low-carbon option for providing heat, but its contribution to airborne $PM_{2.5}$ levels results in the equivalent of about 1,500 premature deaths each year. Again, is that a price worth paying? At present, there is no framework which would enable us to carry out that balancing act, alongside consideration of whether recovering energy is a sensible thing to do with straw, or whether it can be used more productively.

So, there's a few ideas. No more diesels, household wood burners or straw-fired agricultural boilers. Better controls on shipping emissions. Put that lot together, and you have a low-cost way of avoiding almost half the UK's $PM_{2.5}$ emissions, and – more importantly – avoiding many of the associated 30,000 or so premature deaths each year.

$PM_{2.5}$ AROUND THE WORLD

Priorities for reducing emissions of $PM_{2.5}$ are different elsewhere in the world. In many countries, less progress has been made in reducing emissions of particulate matter compared to the improvements secured in the UK. The figure shows how emissions of PM_{10} worldwide changed between 1970 and 2008.

And it shows that they didn't change much. A few ups and downs, and shifts in the countries and activities which give rise to the most PM_{10} emissions, but the net effect of all that is that, by

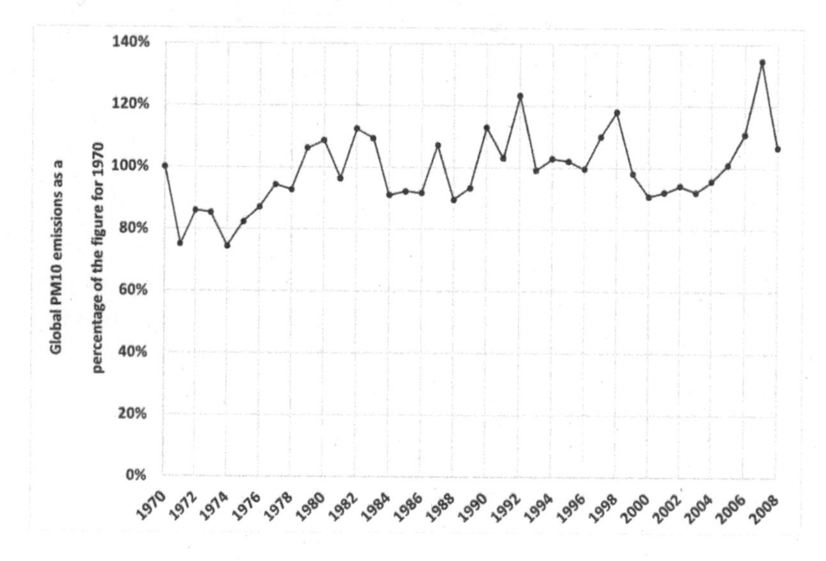

Global emissions of PM_{10} from 1970 to 2008[36]

2008 PM_{10} emissions were pretty much the same as they were in 1970. And the same is probably true of $PM_{2.5}$.

The twelve countries which contributed more than 1% of global $PM_{2.5}$ emissions in 2010 are: United States, Brazil, Nigeria, Ethiopia, South Africa, India, Pakistan, China, Myanmar, Thailand, Vietnam and Indonesia.[36] These countries accounted for more than 70% of global emissions – indeed, China alone was responsible for 30% of the worldwide total. Residential combustion dominated $PM_{2.5}$ emissions from African and Asian countries, accounting for 80% or more of estimated emissions from Nigeria, Ethiopia and Pakistan, and about half of emissions from South Africa, India, China and Vietnam. Agricultural waste burning is also a consistently high source, contributing half of Brazil's and Myanmar's emissions. Industrial sources were more important in the United States, India and China.

This highlights the importance of tackling solid fuel combustion in the home throughout the world, and particularly where people don't have access to less polluting sources of fuel for heating and cooking. The World Health Organization estimates that around 3 billion people burn wood, coal, animal dung and crop wastes on open fires and simple stoves for cooking and heating. As well as its contribution to outdoor air pollution, burning solid fuels in the home has direct health consequences for the people living there. Just sharing a room with an open fire is dangerous, particularly for young children and the elderly, before we even consider the pollution impacts. And what about those impacts? The WHO states that "over 4 million people die prematurely from illness attributable to the household air pollution from cooking with solid fuels."[77] These deaths are caused by pneumonia, stroke, heart disease, obstructive lung disease and lung cancer. Dealing with these direct impacts of smoke in the home, as well as the millions of deaths caused to people in the wider environment, will require the provision of alternative ways for these 3 billion people to cook their meals and heat their houses. It's a long-term project, but doable: right now, clean stoves are being delivered to ten million people in Indonesia. This would bring real health and quality of life benefits for people throughout the world.

Agricultural waste burning also crops up, no pun intended, as an important source of $PM_{2.5}$ in many countries. Burning waste materials is an attractive option for many farmers, but it's by no means necessary. For example, despite the substantial agriculture sector in China, there has in the past been only a minimal contribution from burning of agricultural waste to national $PM_{2.5}$ emissions compared to other countries. Traditionally, crop residues were collected by farming family members, and dried for use as fuel in the home. Leftover materials can also be composted or ploughed back in to return nutrients to the soil. Even in China, however, these more labour-intensive methods are becoming too expensive, and crop burning now makes its annual contribution to pollution.

Alongside the more traditional means of re-using waste agricultural materials, a further option widely available nowadays is anaerobic digestion of agricultural residues. This process results in a liquid digestate residue which can be spread on fields as a fertiliser, together with a biogas product. The biogas can be burnt to generate heat and electricity, or could potentially be added to a local gas network. One option to avoid burning of agricultural wastes may be to develop local anaerobic digestion facilities which can serve a large number of local farmers, and the products used to benefit the farmers and local residents.

PROGRESS REPORT ON URBAN AIR QUALITY

So we can point to success stories for some pollutants, some sources and some regions. We've done a good job of rescuing the ozone layer – at least I hope we have: it'll be a few years before we find out for definite. We've eliminated lead from vehicle emissions, with all the benefits that brings for child development. In the UK and throughout Europe, we've dealt with the choking smogs caused by smoke and sulphur dioxide. Sulphur dioxide remains a problem in other parts of the world. Catalytic convertors have also been great news for air quality, eliminating carbon monoxide as an air quality problem around the world, and doing a lot to reduce levels of nitrogen dioxide and hydrocarbons... and hence, ozone. Problematic as ground-level ozone is, it would be a whole lot worse if we didn't have catalytic convertors on petrol-engined cars.

All that good news means that globally, air pollution is now only responsible for about seven million premature deaths each year. Only. That is, of course, seven million too many, and what's worse is that a lot of those deaths are avoidable. In the next few chapters, we'll look at the tools and methods at our disposal to understand air pollution, and then in the final chapter, we'll look at what the future might hold. Will our efforts to deal with air pollution be like our feeble attempts to reduce emissions of greenhouse gases – very expensive with only marginal success? Or will we be able to reproduce the success of campaigns to get rid of lead in vehicle fuels and CFCs in aerosol cans? Wouldn't it be great if we could consign air pollution to ancient history, the same as we have done with CFCs, smallpox, smoking in restaurants, and driving without a seatbelt, even if it would be terrible for my career? It's a sacrifice I'm prepared to make.

AIR POLLUTION IN THE NEWS

At least people affected by poor air quality seem to know a bit about it nowadays. How do you make a complicated, technical subject like air quality engaging and interesting for a wide audience? Here's an article from *The Sun* from May 2017.[78]

> ***SOMETHING IN THE AIR What is toxic air pollution, what part of London has the worst smog, what does it do to your lungs and should you stay indoors?***
>
> ***AIR pollution can cause serious health problems including stunted lung growth and cancer, experts say.***
>
> *But what causes the toxic air and what is being done to tackle it?*
>
> ***What is toxic air pollution and how is it caused?***
>
> *Hazardous pollutants from industrial facilities can have serious negative health impacts as they are released into the air.*
>
> *The main pollution problem in both developed and rapidly industrialising countries has been high levels of smoke and sulphur dioxide – emitted through the combustion of sulphur-containing fossil fuels such as coal.*
>
> *Petrol and diesel-engine vehicles emit several types of pollutants such as carbon monoxide, oxides of nitrogen, volatile organic compounds, and particulate matter.*

In general, pollution from industry is steady or improving with time, however, the consequences of traffic pollution are worsening with time.

By January 19, 2017, London had already breached its legal limits for toxic air for the entire year.

The quality of air pollution in London is thought to be contributing to about 9,000 deaths every year.

And so *The Sun* goes on, for over 1,000 words, covering some quite technical subjects – what air pollutants do to us, what the air pollution alert levels mean, and what is being done in London to improve air quality. The sort of topics that any reader of *Every Breath* takes in their stride. And *The Sun* is also not afraid to re-use its material: an expanded article was published in July 2018 (this was next to a fascinating article on Meghan Markle's staffing issues when I read it, so some readers may have been understandably distracted). This time, *The Sun* went with a headline asking "how has the heatwave affected health levels?" Because we all like to talk about the weather, don't we?

I could quibble with one or two of the comments and terms used, but by and large, the information provided in these articles is pretty accurate. And there is a surprising level of detail there – some history as well as descriptions of levels of air pollution, what the sources are, and what effects they have on health for Londoners. The journalistic style of very short paragraphs, almost written like bullet points, enables the writers to set down a series of only loosely connected facts and comments. Didn't get that one, or didn't like it? Here's another one. The article goes on to talk about what people can do to avoid the worst effects of air pollution episodes, and to contribute to air quality improvements longer term. Finally, there's more of a political angle, with the views of both the (Conservative) Prime Minister Theresa May, and the (Labour) Mayor of London Sadiq Khan quoted extensively and without comment. Maybe if this piece had been about a less scientific subject, there would have been more of a party political slant to the final section, but as it is, both politicians' views and plans are quoted for the reader to make up their own mind.

The most exciting aspect of this article, even allowing for insights into the Duchess of Sussex's domestic arrangements, is that it's there at all. Just a reminder that this article was in *The Sun*, a right-wing tabloid, not in the left-leaning environmentally focused *Guardian*. *The Sun*, along with all the other mainstream UK media, now carries a new story every few days on air pollution. Time was when I used to buy a copy of a newspaper with a story about air quality, carefully cut out the item, scan it, and circulate to my colleagues in tones of awe that someone, somewhere was interested enough to put a story about our little corner of environmental science in the papers. Nowadays, it's hard to keep up with the consistent stream of new stories about air quality – often describing the latest legal developments, or half-hearted government responses to an air quality problem. A quick search on *The Sun*'s website revealed that there were forty-four separate items on "air quality" in the first four months of 2017 alone, and *The Sun* has made me very happy by describing the newly confirmed end of petrol and diesel cars in the UK by 2040 as "carmaggedon". That's forty-four stories in just four months, in just one tabloid newspaper which doesn't have a particularly strong reputation for banging the drum for the environment. The world has changed.

AIR POLLUTION IN ADVERTISING

Here's another approach to talking about air quality. The Body Shop, a cosmetics retailer, has sponsored the installation of air pollution removal hardware in a number of bus shelters in central London. The advertisements pick up on the word "airpocalypse" which has become prevalent in China to describe severe air pollution episodes. Let's hope we can avoid an airpocalypse with a well-timed carmageddon. The idea is that the bus shelters have been adapted to remove nitrogen dioxide and particulate matter from the atmosphere, making a better micro-climate for people waiting inside the shelter. It's a good way of raising the profile of air quality in a memorable and engaging way, as well as demonstrating the removal technology, and maybe also selling a few of the Body Shop's anti-pollution "city skin" products.

But what does the writing say on the advert? *"This year central London's nitrogen dioxide levels have exceeded the legal limit 99% of the time."* That's a bold claim... but also quite vague about what it's saying, and not particularly plausible if you know about how pollution levels compare to the legal limits for nitrogen dioxide. A quick exchange of emails with Airlabs, the company behind the pollution removal technology, told me what they'd done to calculate this eye-catching number of 99%. It turns out that they have looked at daily average concentrations at five nearby air monitoring stations, and found that over a four month period, 99% of the daily average concentrations were above the air quality standard for annual average concentrations. I checked the maths, and they're quite right about that. But what they've done here is compare shorter averaging period concentrations against the standard for longer averaging periods, as well as looking at less than a year's data and on top of that, combining results from different monitoring stations into a single value. So there are no legal implications from the calculation, despite the use of the term *"legal limit"*. It's certainly true that if you exceed the annual mean concentration for enough days, the chances are that you will end up exceeding the annual mean concentration over the course of a year – and that does have legal implications. But as a slogan to put on a bus stop, *"This year central London's nitrogen dioxide levels look likely to exceed the legal limit"* is pretty feeble compared to the attention-grabbing version that the advertisers went with. And there's a lesson there. There's a lot of competition for our attention, so to make people sit up and take notice of air pollution, you need something a bit more hard-hitting than *"likely to exceed the legal limit"*. So I don't blame the advertisers for stretching the truth here. It certainly made me sit up and take notice, and I'm sure it'll help to raise awareness of air pollution, perhaps among a bus-riding cosmetics-using demographic who are not otherwise strongly targeted by air quality messaging.

AIR POLLUTION CAMPAIGNS

Shock tactics were also used by Clean Air Now, a youth-led campaign which aims to raise awareness of air quality issues among young people. They used a billboard campaign which deliberately highlights the urban focus of air pollution and the fact that young Londoners are growing up in a polluted environment, asking "is this the future we want?"

Clean Air Now campaign poster[79]

A different approach was adopted in a collaborative project led by physical chemist Tony Ryan and linguist Joanna Gavins at the University of Sheffield. For three years, an air pollution eating poem was displayed on the outside of a building at Sheffield University. OK, it's the cloth the poem is printed on which is eating the pollution, not the four-line free metre poetical structure. The university estimates that the cloth has removed over two tonnes of nitrogen dioxide from the atmosphere over this time. That's less than a tonne per year, not a huge quantity compared to the million tonnes emitted from the UK annually, so as with the bus shelters, the value of this initiative is more in putting air pollution into the public eye than in what is actually achieved. And a great poem from Simon Armitage:[80]

> ***In Praise of Air***
> *I write in praise of air. I was six or five*
> *when a conjurer opened my knotted fist*
> *and I held in my palm the whole of the sky.*
> *I've carried it with me ever since.*
>
> *Let air be a major god, its being*
> *and touch, its breast-milk always tilted*
> *to the lips. Both dragonfly and Boeing*
> *dangle in its see-through nothingness...*
>
> *Among the jumbled bric-a-brac I keep*
> *a padlocked treasure-chest of empty space,*
> *and on days when thoughts are fuddled with smog*
> *or civilization crosses the street*
>
> *with a white handkerchief over its mouth*
> *and cars blow kisses to our lips from theirs*
> *I turn the key, throw back the lid, breathe deep.*
> *My first word, everyone's first word, was air.*

The innovative and eye-catching messaging in the Body Shop, Clean Air Now and University of Sheffield initiatives is definitely getting the message that air quality is important out to many people who even a few years ago would not have claimed to have any interest in air quality. More traditional campaigning techniques are also important in bringing the key issues on air quality to the attention of politicians and other decision-makers. For example, the British Lung Foundation organised a petition calling for action to save children's lungs from air pollution, which was handed in to Downing Street in December 2016. But, though it pains me to say it, maybe petitions and reports have had their day, and the future of campaigning and lobbying lies in Twitter and bus shelters.

As well as national campaigning, air quality has occasionally been a local issue for many years – for example, any time a new waste incinerator is proposed, or maybe when a local authority suggests restricting city centre traffic to improve air quality. More tellingly, air quality can become a local concern when a local authority is seen to be avoiding its responsibilities to improve air quality. When a local community realises that they are being exposed to high levels of air pollution, this can trigger anger and a desire for change. An active local pressure group can bring concerns or anger about persistently high levels of pollution to a wider audience, as for example the East End Quality of Life initiative in the Brinsworth and Catcliffe districts of Sheffield has done for a long time.

AIR POLLUTION ON SOCIAL MEDIA

While these local and national campaigns are important, in the past few years, air quality has moved beyond occasional or even long-lasting local difficulties to become part of the mainstream news agenda. There are now proper stories featuring legal cases – notably, the UK being threatened with fines by the European Union for failing to meet air quality standards quickly enough which plays into a wider narrative about Britain's relationship with the EU. This has been backed up by legal action against the UK government brought and won by a small activist group, ClientEarth, which brings a David and Goliath or Robin Hood element to the stories, and you've got to enjoy an underdog sticking it to the big guys. There is big money involved – delivering the kind of improvements that will be needed to comply with the air quality standards will require multi-million pound investments. Indeed, the British government is promoting a *"£3 billion programme to clean up the air and reduce vehicle emissions"* which includes £1.2 billion of investments in cycling and walking up to 2021, and £1 billion for ultra-low emissions vehicles. At the same time, there is a steady stream of new information on major health issues to be reported on – and while many people will turn over the page from a story about the environment, those same people may well stop and read a story about their own or their family's health. And on top of all that, when the Dieselgate story broke, we had a pantomime villain in the very well-known and tangible form of Volkswagen – with over three million VWs on the UK's streets,[81] a reminder of the illegal measures introduced by Volkswagen to defeat emissions tests is never far away. With a steady stream of genuinely significant, interesting, accessible mainstream news stories, air quality has risen relentlessly up the news agenda for the past few years.

To see whether air pollution genuinely is getting more coverage these days, I looked at how many hits I could get by searching on the "News" section of a well-known Internet search engine in December 2018. Here's what I found:

- "Air pollution": nine million results
- "Water pollution": 126,000 results
- "Soil pollution": 8,500 results
- "Noise pollution": 90,000 results
- "Obesity": thirteen million results
- "Passive smoking": 12,000 results
- "Brexit": 210 million results – sure, we are in the middle of debates about the Brexit deal, but even so, that's a lot of news.

- *"I'm a Celebrity"*: twenty million results – similarly, I'm writing in the thick of the 2018 high jinks in the jungle which recently saw Noel Edmonds leave the show and Harry Redknapp emerge victorious, and if that isn't worth twenty million news stories, I don't know what is.

So it looks like air pollution is only half as important to the news agenda as *"I'm a Celebrity"* but I'm very hopeful that it's just a temporary blip, and once Noel, Harry and the rest of them have slipped off the radar, air pollution will still be front page news. It's interesting to see that water, soil and noise pollution generate a lot less news coverage than air pollution. Perhaps that's because we can treat water and soil contamination, and insulate ourselves against noise, in ways which are not available for dealing with air pollution. Unfortunately, I didn't do the same test five years ago, so I can't say how the news coverage picture has changed over that time, but I did run the test eighteen months previously, the day after the UK's general election in June 2017, and there was much less coverage of air pollution even then. Still that's just two days, so definitely not a statistically significant observation.

What about social media? Perhaps surprisingly, as an environmental issue with widespread effects on large numbers of people, air pollution seems to have a relatively low profile on social media. Maybe this is where *"I'm a Celebrity"* gets its own back. One of the key information providers on social media is the World Health Organization: its Twitter account has 3.7 million followers, but of course the WHO provides information covering a very extensive range of health and environmental issues – not just air pollution. Still, a tweet in June 2017 describing air pollution as "the invisible killer" generated 1,400 responses (replies, retweets and likes). The UK government's air quality alert Twitter feed, @DefraUKAir, has a, perhaps disappointing, seven thousand followers – about one in ten thousand members of the UK population. ClientEarth, the legal activitism group, tweets actively about air pollution to its twenty-three thousand followers. Other groups focusing on air quality such as Healthy Air, Clean Air in London and LondonAir (yes, they are different) have managed to attract up to thirty-five thousand followers. And what of myself? As of now, @BroomfieldAQ can point to very nearly forty dedicated followers. I like to think that it's not quantity, it's quality, even though that's demonstrably untrue in the world of social media.

The well-established environmental campaign groups have hundreds of thousands of followers for their UK operations, and do provide information on air quality issues although this tends to be a bit lost amid the much more extensive tweeting about climate change. And of course there are professionals and concerned individuals circulating information on air quality. This is all great, but to be honest, it doesn't feel like a social media-led popular groundswell likely to lead to concrete or lasting change in how we deal with air pollution. There is more a feeling of a limited network of interested people and activists talking to each other. The attention paid to air quality through the mainstream media is probably more important in determining the influence that air quality impacts have on national policy and public awareness. Air quality is genuinely an important, interesting, shocking, relevant topic, and that gives it a life of its own, enough to make its own way in a crowded news agenda. Which makes this a great time to know a bit about the air that we breathe: as with anything else, as soon as you know a little of the background to a news story, you can start to sniff out the half-truths, spin, and indeed the downright lies.

REAL DATA

One of the advantages of investigating air quality now is the opportunity to take advantage of the wealth of information which is available to anyone with an Internet connection. There are plenty of opinions out there on air pollution: whether it matters, who's responsible, what needs to be done and who should pay.

But leaving aside opinions, there is also unparalleled hard data on air quality available to anyone who wants it. The UK's air quality archive[82] contains all the data recorded at over 1,500 sites from 1961 to yesterday. Another five and a half million measurements are collected every year,[83] so there's plenty of data there – it's where I got hold of the monitoring data which I dissect in Chapter 9. And that's not all: more data is available from Scottish, Welsh and Northern Irish databases, regional resources and individual local authority websites. For countries in the European Union, including the UK at the time of writing, air quality monitoring records are stored in the European Environment Agency's Airbase system, and many countries have their own national databases of course. I had a quick look at the data available for Paris and checked that yes, you can indeed download air quality data at sites including La Defense in the city centre. The levels measured there were pretty comparable to those measured at L'Élephant et Chateau in central London, and reported via the LondonAir website. There's such a wealth of freely available information out there, and it's a bit of a shame that we don't do more to make the most of it. It could tell us about sources of pollution in local communities, and whether things are getting better or worse. It can tell us about the characteristics of air pollution episodes, and what we need to do to address the short-lived peaks in levels of air pollutants, alongside making improvements in longer-term average levels. You could do an air quality check as part of the process of buying a house or choosing a school for your children. Of course, air quality isn't likely to be the biggest reason for taking these fundamentally important decisions, but it might be worth – say – 14% or so.

IT'S GOOD TO TALK

Just occasionally, I'm allowed out of the office to talk to people face to face about air quality. This happens most frequently in relation to new development proposals – particularly, you won't be surprised to hear, waste to energy plants. These discussions are most effective when people who are worried about a new proposal can talk face to face about their perspective and their worries. If people are worried about the effect on their house price or the physical appearance of the new facility, there is little that I can do to help. But often, people have heard that a new waste incinerator is going to be built, and that it will cause cancer or harm babies in the communities living nearby – and that's a genuine, heartfelt, sickening worry.

In those one-to-one conversations, the first thing to do is listen to what people have to say. I've generally found specialists like me to be very keen to start talking about whatever it is that we know about, whereas it's always better to start by shutting up and listening. Any individual might be angry, afraid, confused, cynical or even supportive, and if we don't listen carefully, we won't be able to engage in a conversation which will be useful and productive for everyone. Once we have understood the concerns that people have, we can perhaps start to talk through how we deal with them. Local communities can easily get the impression that a new development is going to get the go-ahead without any consideration of the air pollution or health consequences. The word "reckless" is often used to describe waste incineration proposals, but it's a completely inappropriate description of the work that we do. We make sure that the air quality and health impacts of a new proposal are carefully evaluated, and further reduced where necessary below the minimal levels which result from conforming with the legal limits on emissions from waste incineration in Europe. It's anything but reckless. Talking through with people how we carry out these studies, taking account of the local environment, is often an eye-opener for all concerned. I get to find out features of the locality that I might not otherwise have been aware of. And I often find that local residents, even if they don't like or accept what I'm saying, can appreciate that air quality and health risks are being evaluated, and that the studies are being carried out by human beings who are willing to listen and talk openly about the details of their work.

Face-to-face discussions can be useful for everyone, whereas trying to address a large-scale meeting is often thoroughly counter-productive – although sometimes unavoidable. On a couple of occasions, I have been quite relieved to get away from a meeting unscathed where there has been genuine anger at the message I and colleagues have been presenting. The age of the "man in the white coat" or "trust me, I'm a doctor" approach to science communications is well and truly over, thank goodness. So, standing at the front of a hall and talking at people, even if you've got a really watertight PowerPoint presentation, sets the wrong tone and people quickly and probably rightly start objecting to being preached at.

Furthermore, the air quality and health issues are too complex to be summarised in short soundbites, which is often all that counts in a large meeting. If I say something like, "The proposed waste incinerator will be safe," this immediately triggers all sorts of comments, mostly unprintable and some even justifiable. But a more nuanced comment like "Our air quality assessment shows that the proposed waste to energy plant will not have any significant adverse effects on air quality or on public health consistent with the findings of relevant epidemiological studies" doesn't work either – I would probably be booed off before I got halfway through. Indeed, I have been booed and almost chased out of public meetings in London, Surrey and Yorkshire – not because what I was saying was wrong, but because I tried to say it in the wrong way. On the other hand, I particularly remember one lady questioning me in detail for a long time, and ending up saying, "Well, I still don't like it, but I can't think of any more reasons why. Thank you." It was the "thank you" which reassured me that British politeness is still alive and well, and it's something that no amount of written reports or PowerPoint presentations could buy. No, give me the opportunity for face-to-face chats every time.

In this kind of conversation, I've found that we can go into quite a lot of detail, and deal honestly with some of the nuances and uncertainties which tend to get lost, both in large-scale meetings and (dare I suggest it) in 140, or even 280, character tweets. I guess it's easy for me to enter into these discussions, where I'm on home territory and have little to lose. But when you have to explain something to someone who's not familiar with the subject, that's when you find out exactly how well you do understand it. These conversations have sharpened my thinking about some aspects of the work – in particular, how we deal with local weather conditions, and how we address the emissions from an industrial or waste installation when things are going wrong. And most people, including myself, go away from these conversations thoughtfully. By no means does everyone agree with me, or change their mind when they hear what I have to say. Nor do I change my mind, very often. But people do leave with the understanding that there is an alternative point of view, even if they don't like it or espouse it. And there is also the personal contact aspect: a detailed conversation about something that concerns both of us means that we both leave with an appreciation that we've been dealing with another person. Not some mad scientist, incompetent buffoon, sharp-suited operator or hired gun in my case; and not some prejudiced or blinkered objector in their case. At least, that seems to happen a lot of the time, even if I can't honestly say that there is an outcome of mutual respect after every single conversation. And I would have to admit that it's a small-scale, reactive process – important where it's needed, but not really an engine of wide scale social change. We do need change on the grand scale to make substantial improvements in air quality for everyone.

There is a long and honourable tradition of presenting important scientific ideas through comedy and performance. At the 2017 Edinburgh Fringe festival, for example, there were almost forty shows with a science theme, all the way from exploding elephant toothpaste to anti-gravity and Schrödinger's cat. I tried to write a stand-up routine on air pollution, pulling out some of my killer jokes, but to be honest, I couldn't really see one-liners about Antoine Lavoisier working in a sweaty upstairs room in an Edinburgh pub. The world will have to go on waiting for the first ever air quality gag-a-thon.

So, the future of telling people about air pollution isn't in stand-up comedy, and nor is it likely to be several billion individual tête-à-têtes (or should that be têtes-à-têtes). Looks like I'll have to write a book or something.

CHAPTER 7

YOUR STREET, YOUR NEIGHBOURS, YOUR HOME

Within A Kilometre

Local air quality has been the backdrop to my life. The first step in dealing with air quality problems is to identify all the sources that contribute to air pollution, particularly in areas where it's bad, which usually, though not always, means busy road junctions and congested streets.

THE DYNAMIC ATMOSPHERE

One of the fascinating features of the atmosphere is how varied and dynamic a place it is. You only have to look out of the window, or even go outside if you're feeling brave, and you can experience all kinds of variations in the atmosphere – sun, wind, rain, snow, clouds, fog, thunder, lightning, rainbows and St Elmo's fire for example. While camping in Wales, I'm pretty sure I've experienced all of them at once. Experiencing a glorious sunset is one of nature's iconic experiences – made all the more exhilarating by its ever-changing nature. You can't watch the same sunset twice. And the interaction of light and cloud changes all the time as the sun sinks closer to the horizon, and the sun's rays have to travel through more and more of the atmosphere.

The movements of air within the atmosphere are a key influence on the air pollution that any individual experiences. It's hardly a surprise to learn that you can experience fresh(er) air by moving away from a source: going from the city into the country, for example – well, as a rough rule of thumb you can. But there are plenty of other factors at work, which mean that the highest levels of pollution aren't always close to the source.

For one thing, one of the reasons that we have chimneys is to enable air pollutants to disperse away from the immediate vicinity of a source. So a chimney doesn't do anything to reduce the quantity of pollutants emitted, but it spreads them over a wider area so that no one has to put up with excessive levels of the released substances. Because of the dispersion processes, the highest levels of pollutants from a chimney are likely to occur about five to ten chimney heights downwind of a chimney, as a rough rule of thumb. So if you live near a factory with a thirty metre chimney, for example, you might expect the highest ground-level concentrations of whatever's coming out of the chimney to occur about 150 to 300 metres downwind.

As well as dispersion from elevated source, it takes time for some pollutants to form in the atmosphere. Ozone, for example, is formed in the atmosphere by the reactions of other substances, mainly emitted from road traffic – it isn't directly released into the atmosphere (as we saw in Chapter 4). Because these reactions take time to proceed, the highest levels of ozone often occur miles away from the traffic and other activities giving rise to the precursors. So if you are particularly sensitive to ozone, about the worst place to live is in the countryside downwind of a large urban area.

PREVAILING WINDS

In the UK, the prevailing wind direction is from the west and south-west. So, more often than not, winds in the UK come in from the Atlantic and blow air pollution, dust, smells, litter, and carelessly-worn hats towards the east and north-east. This has some important consequences, not only for air pollution, but also for the value of your house. Yes, we have finally come pretty much to your front door. The air that's right in front of, and imminently in, your nose.

Before we get into house prices, here's a cautionary note: when you spend as much time as I do thinking about prevailing wind directions, it's easy to forget that the wind doesn't always come from the prevailing direction: it just comes from this direction more often than not. The wind actually comes from pretty much all directions for some of the time – particularly in a windy island like Great Britain. For example, here is a wind rose recorded at Birmingham Airport. between 2005 and 2017. The line in the graph shows how much of the time the wind was blowing from that direction.

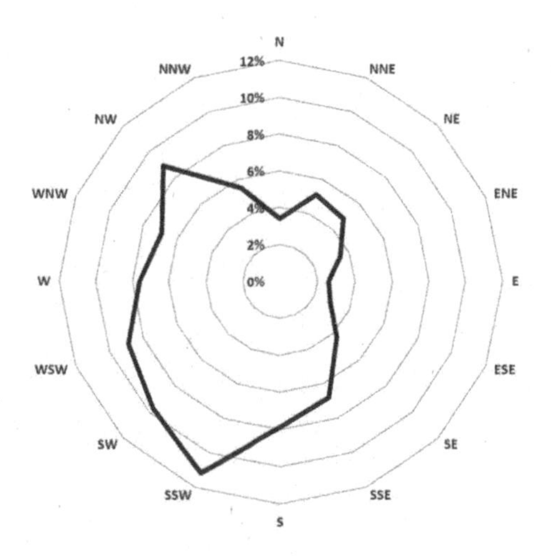

Wind rose recorded at Birmingham Airport showing % of time wind blew from each direction[84]

This wind rose shows that the most common wind directions were between west-south-west and south-south-west – no surprise to anyone who's been outdoors in Great Britain. But as well as the prevailing south-westerly winds, there is also quite a strong prevalence of north-westerly winds, and the suggestion of a reasonable frequency of winds from the north-east. Those winds can come from anywhere. This pattern of prevailing but not exclusively south-westerly winds is pretty typical of most of the UK, although there can be local variations – for example, near the coast, winds are likely to be influenced by onshore and offshore winds. Also, hills and mountains can strongly influence wind flows – for example, channelling winds along a valley. Here's the wind rose for Edinburgh Airport from 2000 to 2017, which is influenced by its location in an east-west running valley. Winds from the west-south-west and north-east are much more prevalent, whereas there is next to no contribution from winds to the north-west and south-east.

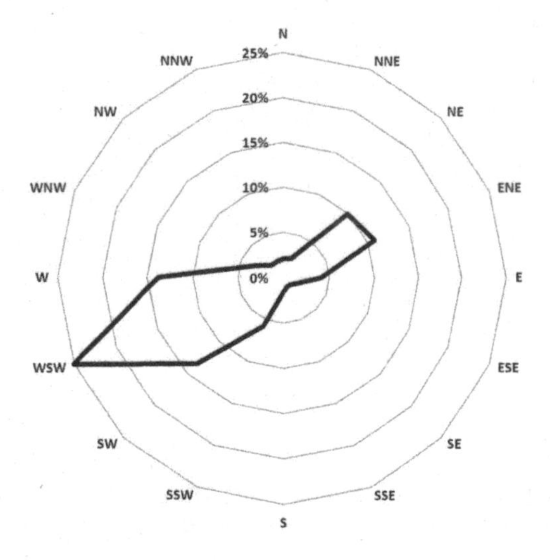

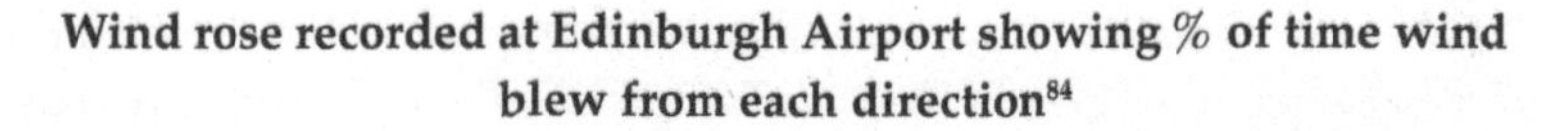
Wind rose recorded at Edinburgh Airport showing % of time wind blew from each direction[84]

You can often see physical evidence for prevailing wind directions in the shapes of trees which have been bent from years of exposure to strong winds from one dominant direction. This can be a bit misleading when thinking about the effects on pollution, because the effects on trees and bushes is likely to be dominated by the strongest winds. Any self-respecting tree won't be too bothered by gentle breezes, even if they do all come from one direction. In contrast, when we're dealing with air pollution, it's often the gentle breezes that result in the worst pollution.

All these winds, prevailing or otherwise, take place in the troposphere – the very lowest bit of the atmosphere. In fact, we often distinguish two parts to the troposphere, this time based not on temperature profiles but on the interactions between the atmosphere and the Earth's surface. The planetary boundary layer is the very lowest part of the atmosphere. This is the part of the atmosphere which is directly affected by the Earth's surface, so there is a correlation between wind speed and height above ground. Higher up within the boundary layer, wind speeds are generally faster than closer to ground level. Because of the friction introduced by surface features such as trees and buildings, wind speeds go all the way down essentially to zero when you are adjacent to the ground. By "adjacent", I really do mean that you have to be right down at ground level to experience no wind at all – the height of a blade of grass or a pebble.

THE IMPORTANCE OF BUMPINESS

Not surprisingly, the more bumpy the Earth's surface is, the greater is its effect on wind speeds. The sea presents much less resistance to wind flows than, say, Manhattan does. Which goes some way to explaining why there are more wind turbines in the North Sea than there are on the sidewalks of New York. We have to take account of this effect when carrying out air quality modelling studies. The increase in wind speed with height can be useful: a relatively small increase in the height of a chimney can deliver a decent benefit in reducing ground-level concentrations of released substances. Apart from anything else, concentrations of substances emitted from chimneys are roughly inversely proportional to wind speed, and pushing the stack height up a bit effectively increases the wind speed at the point of release, thereby reducing concentrations.

As well as slowing the wind down, surface features also introduce vertical movements into the atmosphere, which contribute to the dispersion of emissions. For ground level sources, like road vehicles, a bit of vertical dispersion is useful, because it spreads out emissions upwards into the atmosphere where they can do less harm. That results in lower concentrations at ground level, where you and any children or elderly relatives are typically located. However, for elevated sources, like factory or power station chimneys, a lot of vertical movement in the atmosphere tends to bring released substances to ground level more quickly. So the presence of tall buildings and trees can result in higher concentrations close to elevated point sources than would be the case where the landscape is more uniform.

CYCLONES AND ANTI-CYCLONES

Above the boundary layer, in the upper part of the troposphere, wind flows are driven mainly by pressure gradients in the atmosphere. We're familiar with charts showing areas of low and high pressure on the daily weather forecast. Where there are strong areas of low or high pressure, and the isobars on the weather map are close together, that means that we can expect strong winds. An area of low pressure is caused by warm air at low level rising upwards. This leads to air flowing into the area of low pressure from all sides, and as it flows inwards and upwards, it rotates anticlockwise. This anticlockwise spin is caused by the rotation of the Earth, and works in the opposite sense in the southern hemisphere.

So as a succession of areas of low pressure comes towards Europe across the Atlantic – these are the rather feeble remnants of cyclones from the Caribbean – the air flows around these lows travel in an anticlockwise direction. These anticlockwise airflows usually translate into westerly and south-westerly winds, as the cyclones approach the UK.

The reverse process happens when we experience areas of high pressure, which are also known as "anticyclones" as they are more or less the opposite of the cyclonic areas of low pressure. The airflows around an anticyclone move in a clockwise direction (whereas the winds around a cyclone move anticlockwise. What's confusing about that?). What that means for wind directions depends on where you are relative to the anticyclone. From the perspective of the UK, if you're on the west side of the anticyclone, you can enjoy balmy, warm southerly winds. Conversely, if you're on the east side, you're likely to experience much colder winds from the north.

While high pressure is often responsible for warm summer days and cold, clear winter days, it's the low pressure areas which

Farmhouse upwind of its farmyard[85]

we experience most often out here on our lonely island. That means that we experience winds from the west or south-west more frequently than other directions. And these prevailing wind directions are important for the value of your house. It's no surprise that people have been very well aware of which way the wind blew in their neighbourhood since time immemorial. This local knowledge is built in to the design of farms, which tend to put the farmhouse on the south and west side of the farmyard, and the smelly cowsheds, pigsties and manure heaps on the far, downwind side. I took a look at the closest farm to my house, and (leaving aside the fact that they've built a golf driving range on the farm), sure enough the house is on the south-west side of the farmyard. I was particularly happy to see that the golfers, in contrast, are directly downwind of the farm buildings.

AIR POLLUTION AND THE PRICE OF YOUR HOUSE

Before there was a lot of industrial pollution, the best places to live were usually in the centre of towns. Westminster in London, the Royal Crescent in Bath, the New Town in Edinburgh – these were the addresses where the kings, courtiers and higher echelons of society would live. However, when the industrial revolution came along, the nouveau riche bosses were quick to pick up on the principles of prevailing winds which farmers had known about for centuries. A lot of industrial processes were smelly and smoky, not to mention dangerous. The entrepreneurs and managers of these processes did not want to spend their lives and bring up their families in the miasma of their own operations. So, with commendable acumen and self-preservation, they built themselves big, posh houses on the upwind side of towns and cities. The city centre and downwind areas were, by and large, left for the workforce who didn't have much choice about where they lived, and in any case needed to be near to their places of work. In my own town, back in the twelfth century, a leper hospital was founded right on the eastern edge of the town. There aren't any records, but I do wonder if the monks put it there so that the prevailing winds would take what they would have thought of as the miasma from the hospital away from the city.

Nowadays, our cities aren't generally full of large-scale polluting industries or leprosy hospitals – the factories have been moved out of site to edge-of-town industrial estates, or more often than not, closed down altogether. Nevertheless, old habits die hard, and big houses last for a long time. The industrial revolution continues to have a long-term effect on the social structure of our towns and cities, which is reflected to this day in house prices. There are, naturally enough, many factors which affect the price of a house – and, as we know, the top three are location, location and location. "Location" might include availability of car parking, proximity to good schools, the ability to walk into town, how far the nearest park / countryside / shop / pub is (an adjacent pub may be a positive or a negative) and many more tangible and intangible factors. And one of the factors, which accounts for about one seventh of the house price to this day, is "location relative to the prevailing wind".

To investigate whether this genuinely remains a real issue today, I checked out the average property selling prices between October 2016 and March 2017 in a number of English towns.[86] I chose a number of towns which I think of as "industrial", to the extent that they had experienced growth in industrial activity and population during the eighteenth and nineteenth centuries. The towns were: Newcastle, Halifax, Derby, Manchester, Northampton, Huddersfield and Bolton. I also included three towns which were well established before the industrial revolution: Cambridge, Winchester and Shrewsbury. And finally, I chose Milton Keynes to represent a post-industrial new town. Then for each town, I looked at the postcode map to identify postcode districts which are west and south-west of the city centre (these are upwind of the prevailing wind direction), and districts which are east and north-east of the city centre (that is, downwind of the prevailing wind direction). Just to keep a personal perspective, my family and I have lived in two of the upwind postcodes, and one of the downwind postcodes.

For my "control" town of Milton Keynes, the average house price in the upwind postcode districts (£312,000) was just a bit higher than in the downwind districts (£293,000) – a difference of 6%. So in a town which was designed and built in the post-industrial era (construction started in 1967), house prices in upwind districts are close to those downwind of the city. Of all the other towns I looked at, only Manchester had higher house prices in the downwind districts than the upwind districts. In the remaining nine towns, house prices were higher in the upwind districts – from 5% higher in Newcastle to over 50% higher in the case of Cambridge. The median price difference was 14%, so the sale price of houses in the upwind postcodes were about 114% of those in the downwind areas.

Maybe it's a bit of a stretch to say that this difference is all due to the prevailing wind. But I'm not sure what else it could be. A report in *The Independent* newpaper suggested that odours continue to have a significant effect on house prices, with a pleasant odour adding about 5% to the value of a house in London.[87] What's perhaps more surprising than the 5% bonus for a nice smell is the penalty for a bad smell: the article quotes Dr Alex Rhys-Taylor of Goldsmiths College as suggesting that an unpleasant odour could take away almost half the value of a house. The article goes on to confirm that "*Since the classical era this has usually been the case, so that west-prevailing winds would keep one side fragrant and the other odiferous. This aspect of urban planning goes back at least to the tanneries and composts of ancient Rome, says Rhys-Taylor, and has persevered in the sense that east ends – in the northern hemisphere at least – tend to be poorer.*" My rudimentary analysis of house prices bears out the view that east ends tend to be poorer, and I've gone ahead and put my (unwillingness to spend) money where my mouth is. Now that I'm no longer directly concerned about school catchment areas, I live on the east side of the town centre, where, according to the prevailing wind theory, there are bargains to be had. Following this cunning plan, we were, perhaps, able to make a 14% saving on the price of our house, which might explain why it was so surprisingly affordable, at least until we start digging under the patio.

SMELLS IN THE PAST

If you were strolling through any large city in the nineteenth century, upwind, downwind or in the middle, I'm pretty sure that the first thing that would strike you would be the smell. Sure, there would be the smells of hot food that we remain familiar with to this day, even if it wasn't yet the era of the Big Mac. But what a mix of other smells: small-scale industries like tanneries and butcheries, horse manure and sewage. Not to mention the combined effect of millions of armpits before the invention of the roll-on deodorant.

By the end of the nineteenth century, for example, New York's horses produced over a thousand tonnes of manure a day, every day, inconveniently spread around the city's streets. In 1894, *The Times* newspaper confidently predicted that "in 50 years, every street in London will be buried under nine feet of manure."[88] Well, in the end, the horse manure problem (and the associated public health problem of dead horses) was solved by the arrival of the internal combustion engine, and motorised transport.

SMELLS FROM SEWAGE

But what about sewage treatment? By the nineteenth century, the most pressing air pollution problem was odours from sewage. Sewage treatment can sometimes cause odour problems today, but that's nothing to the Great Stink of 1858. The problem was that London's sewage was not so much treated as dumped in the river Thames. As London grew, the river was just not big enough to handle all the sewage that ended up there. Leaving aside the risks of taking your drinking water from the same place that you send your poo, by the middle of the nineteenth century, the river Thames had such an unbearable smell that it could no longer be ignored. The Great Stink had been waiting to happen for decades, and in the summer of 1858, low flows in the river Thames combined with warm weather to produce smells so awful that finally some action was taken. It was fortuitous that Parliament had recently moved into its swanky new office building right beside the Thames, where the smells were at their most unbearable. Could it be that the direct effects of the smell on the politicians and civil servants themselves was what spurred the government into at last taking the radical steps that were needed to deal with London's sewage?

The engineer responsible for designing and constructing a sewerage network for London was Joseph Bazalgette, and he was the right man to get the job done, and done fast. By 1865, just seven years after the decision was taken to go ahead, sewers had been constructed across London and the system was opened by the Prince of Wales. Seven years! Works continued to complete the network to transfer sewage downstream, with later upgrades to treat the sewage before discharging the treated effluent into the river. Huge investments were made in sewage collection and treatment in cities throughout the UK in the second half of the nineteenth century, which we are still benefitting from to this day. Bazalgette was a great and visionary engineer, whose designs for the London sewerage system continue to be more or less fit for purpose 150 years later. He built six main interceptor sewers, together with thousands of miles of main and local sewers, which made use of the city's network of buried rivers to transfer sewage to a discharge point, and later treatment facilities, in east London. One of the interceptor sewers is built into the Chelsea and Victoria Embankments, which run along the north side of the river Thames, and where you can stroll today to enjoy the sights of the river Thames, the London Eye, Cleopatra's Needle, and traffic on the A3211 dual carriageway. So if there's occasionally a problem with odours from sewage systems, at least it's not a Great Stink, hopefully, and let's not forget that the infrastructure might well be over 100 years old.

WHAT IS A SMELL?

It's in the nature of odours that they are only a problem if they are detected by someone's nose. There may be a philosophical discussion to be had, along the lines of the famous question about the noise of a falling tree – if an odour forms in a forest where there's no one to smell it, is it an odour?

In contrast, a lot of the pollution we've been thinking about has to be taken on trust. An instrument might tell us how much of one or two pollutants are present in the atmosphere at a particular place and time. A computer model can be used to show pollution levels across a wider area, and to look at what might happen in the future depending on the actions taken to improve air quality. But you can only rarely see or sense air pollution yourself, and even then, only when things are going wrong and pollution levels are high.

The one time that air pollution is clearly detectable is when there's a smell. Localised smells are a fact of everyday life: as we used to say when I was at school in the 1980s, the one who smelt it, dealt it (by the way, as an experienced air quality professional, I can now confirm that's not necessarily a foolproof source attribution strategy). On a more positive note, my favourite smells of freshly made bread, brewing coffee and sizzling sausages always put me in a good mood. Which maybe says more about what motivates me than the particular merits of these smells. Other people may be inspired by the fragrance of roses or freesias, newly mown grass, a fine wine, Chanel No.5, tobacco smoke, or chocolate.

The right smell in the right place can be a pleasure, and a pleasant smell can trigger memories and associations in a similar way to a familiar piece of music. On the other hand, and it's a big hand, the wrong smell, or even a pleasant smell at the wrong time or place, can be unendurable. At the age of twenty-one, I spent a year working in a night shelter for homeless people, and occasionally the distinctive smell of cannabis would permeate the old church building where we were based. Whatever users may say about the relaxing effects of cannabis, I always found that it led to trouble at the night shelter. To the extent that, even now, when the age of twenty-one is a distant memory, the smell of cannabis immediately brings the associations of stress and aggravation into my mind. Cannabis is absolutely not a relaxant as far as I'm concerned. Even my favourite food-related smells are objectionable if I'm feeling ill. I used to live in York at a time when it had two chocolate factories. The occasional smell of chocolate of an evening was rather pleasant. But sometimes we had the same smell first thing in the morning, when I really wasn't in the mood for it, and it was sickly and anything but pleasant.

So, what is a smell? A smell is caused by the presence of volatile or semi-volatile chemicals in the atmosphere. These chemicals stimulate specialised sensory cells in the nose, which are called olfactory sensory neurons, and which each have an odour receptor. When an odour is detected, the neurons send a message to the brain, and the brain analyses the combination of receptor messages to identify the odour.

As well as enabling us to appreciate pleasant smells, the ability to detect odour is an essential survival tool. We instinctively avoid things that smell unpleasant, and it's no coincidence that unpleasant smells are often associated with substances that could be harmful. The pleasantness of an odour can be characterised on a scale from +4 (most pleasant odours) to -4 (most unpleasant odours). The technical term for pleasantness of an odour is "hedonic tone", but there is little apparent hedonism once you get into smells with negative values. The most unpleasant reported odours are something of an axis of evil: those reported as being between -3 and -4 on the hedonic tone scale include smells described as cadaverous (dead animal), putrid, foul, decayed, faecal, vomit, urine, rancid, and reminiscent of sewage, vomit, urine and burnt rubber.[89] There is no doubt that anything with these smells should be given a very wide berth. Our instinctive response is to draw back from all these smells, and this instinct protects us from coming into contact with decaying or faecal matter, and all its associated risks of infection.

Odour detection is a positive survival tool as well. In the animal kingdom, smell is used by predators to detect their prey, and by prey to detect their predators. We've all seen footage from the African plains of a gazelle stopping eating as it detects the faintest whiff of lion... before thinking, "it's probably nothing," and going back to munching some delicious thorns. Nowadays, of course, we hunt our prey using Ocado which doesn't normally require the use of a sense of smell.

Normally, we accept odours as part and parcel of everyday life. In the home, we expect to experience and enjoy the smell of cooking, as well as fresh flowers, freshly mown grass and all kinds of toiletries. We also expect and accept, without necessarily enjoying, the smell of sweaty footwear, teenage bedrooms, and babies' nappies – perhaps at different stages in life. Environmental odours are also with us every day. In the city, we might encounter fast-food odours, or maybe smells from drains, rubbish lorries or other people's bodies on a crowded train. In the countryside, odours are more likely to arise from farming. Slurry spreading is a particularly effective way of spreading odours from a very smelly material into the atmosphere in order to affect the widest possible area. Still, it's an agricultural activity, which enables farmers to make good use of readily available organic material, and one which may occasionally be acceptable in an agricultural setting. Animal housings and manure heaps can also be more persistent local sources of odour.

WHEN A SMELL BECOMES A PROBLEM

Odour problems arise most commonly in relation to cooking smells in urban areas, sewage works, composting and landfill sites. The Chartered Institute of Environmental Health estimated that about 19,000 complaints of odour were received by local authorities in England in 2011.[90] This corresponds to about 400 complaints per million people. Not a huge problem in overall terms, but anyone who's been affected by a pervasive odour will know that it can have a major impact on quality of life.

Cooking odours, particularly from fast-food restaurants, can cause problems because they are often located in residential areas, cheek-by-jowl with houses and apartments. If you add in the possibility of poorly designed and located vents (for example, just below someone's bedroom window is not ideal), and the potential for ventilation systems to become dirtied with food residues and grease, then odour problems may not be too far off.

Questionable location of kitchen flue[91]

Sewage works are the most widespread source of odour complaints. Most sewage works go quietly about their business, without causing any odour problems. Indeed, many people aren't aware of where their local sewage works is (if they have one), which is a good result for the operator and the community as a whole. However, sometimes things do go wrong, and a nasty smell can be the result. Sewage smells are sometimes caused by the sewage arriving at a sewage works: for example, if flows are low so the sewage takes a long time to get to the treatment works, high levels of odorous chemicals can build up in the sewage. These can be released on arrival at a sewage works or pumping station. It then becomes the operator's problem to deal with the odours as efficiently as possible, while also sometimes getting the blame for smells coming from the sewage pipes upstream which have nothing to do with what goes on at the sewage works. However, sometimes smells are caused by a sewage works, particularly when it is not designed or operated well enough to prevent them. The most common problems are caused by difficulties in managing sewage sludge. Sludge needs to be processed quickly, and not left in settlement tanks or overflow tanks, where it can start to decompose with the formation of extremely smelly sulphides.

Once a site has caused an odour for local people, its problems are only just beginning. That's because our experience of odours doesn't only depend on how strong the odour is. Many other factors come into play. There are all the characteristics of the odour – its hedonic tone (pretty nasty, in the case of sewage odours), frequency, intensity and duration. Then there are factors relating to the people experiencing the odour – an individual's sensitivity to the smell, and whether the odours are predictable or manageable.

And then there are less tangible factors, such as how the person affected by the odour perceives the source. You might feel that the activity giving rise to the smell is providing you with a valuable service, and that those responsible listen to your concerns and act on them. Under these circumstances, you're likely to be reasonably tolerant of future odours, as long as you retain confidence in the ability of the operators to keep on top of the odour problems.

In contrast, people are much less likely to be sympathetic to a source of odours which they feel is not giving them any benefit, or where their concerns are not listened to, or if they don't have any confidence in the operator's ability to manage odours. People are also disturbed by odours which are unfamiliar, and where they don't know if the chemicals causing the odour could be affecting their families or themselves in other ways. We find operations which don't have any local benefits particularly unwelcome – for example, who wants to live near a smelly waste facility which handles materials brought in from some distance away? Under these circumstances, however much the facility improves in terms of odours, those living nearby are likely to be highly intolerant of any smell at all. Even a smell which would normally not cause any problems could affect people very badly, and trigger a strong response.

MEASURING SMELLS

So dealing with odours is a strange combination of science, engineering and psychology. And all the more so because of the Heath Robinson approach that we have to take to measuring odours. The problem with measuring environmental odours is that they are usually caused by a complex mix of chemicals. We know quite a lot about the chemicals which cause environmental odours – for example, the rotten egg smell of decomposing eggs and vegetables is due to hydrogen sulphide, together with maybe some other organic sulphur compounds such as dimethyl sulphide and methyl mercaptan. I'm pretty confident on those, because I did my PhD research on the reactions of dimethyl sulphide and methyl mercaptan, which made me the least popular member of the chemistry department for quite a long time. At least, I think it was the chemicals that were the problem. The trouble was, these chemicals are pretty volatile, so there was always a kind of low background level of release from my experimental equipment. As my laboratory was located next door to the technicians' tea room, I soon lost the sympathy of the technical support personnel, and once you've done that, it's game over. And every so often, I would turn a tap the wrong way, or forget to top up the liquid nitrogen in the cold trap, and everyone knew all about it. So, if you were working in or near the chemistry department at York University between 1989 and 1992, can I take this opportunity to say sorry, and I hope the smell has come out of your clothes.

Anyway, sulphides are among the most common contributors to environmental odours. Similarly, we know that faecal odours are largely due to aromatic amine compounds such as skatole and indole (In this context, "aromatic" does not mean "sweet-smelling". Anything but. The chemical meaning of the term "aromatic" is much more prosaic: it means a chemical which contains one or more benzene rings, and this bracket is hardly the place to explain what a benzene ring is). Fishy odours, on the other hand, are due to aliphatic amine compounds such as trimethylamine (and "aliphatic" simply means "not aromatic" – that is, a chemical which doesn't contain benzene rings).

As well as knowing what the principal chemical contributors to some smells are, we have measured how smelly many commonly occurring chemicals are, at least to the extent that we know roughly what concentration of a chemical in the atmosphere could be detected or recognised by half the population. It's not a great measurement, because every individual is different, and a lot of these measured "odour threshold" concentrations vary widely by factors of tens, hundreds or even thousands. But it's the best we've got.

Armed with all this information, it would be nice to think that we could analyse environmental odours by measuring the levels of odorous chemicals. We could measure the levels of each odorous chemical in the atmosphere, work out how strong the odour due to each chemical would be, and so calculate how strong the odours resulting from the mix of chemicals would be. Ah, if only life were so simple for the poor hardworking air quality boffin. For a few sources of odour, we can do this. For example, in the case of a chemical process emitting a single chemical, we can probably make a reasonable assessment of its odour impact by measuring or modelling levels of the released chemical in the atmosphere. In cases where a sewage works odour is characterised by the rotten eggs smell of hydrogen sulphide, we can sometimes assess and monitor the odour by measuring hydrogen sulphide. But that's about it.

For every other type of odour, if we try to assess the odour intensity by adding up the contribution from individual chemicals, we invariably end up not being able to account for most of the odours detected in practice. This is firstly because we probably don't know what all the chemicals contributing to a complex odour are – for example, the smells from a food processing factory, farm, composting site or sewage works are made up of a mix of chemicals, and it's a mix which is never quite the same twice. So we're always going to come up short when we try to work out how smelly the air is from information on its chemical composition. Secondly, these measurements are difficult and expensive to do well, and as a result, the available information on odour thresholds of individual chemicals is not necessarily of the highest quality, and sometimes can be charitably described as pants. And finally, odours are not just caused by the contribution from individual chemicals added up – our noses are a lot more sophisticated than that. Instead, our experience of odours is a response to the complex mix of chemicals presented to the sensory cells and odour receptors in our noses.

This means that the only reliable way of detecting odours is to use the human nose as the detector. Even though everyone's nose responds differently to odours, and our experience of odours is strongly affected by factors other than what the actual smell is – it's the best we've got. So much so that measurement of odours using human noses – or "olfactometry" as we prefer to say; it sounds so much more professional than "sniffing" – is the basis for CEN Standard EN 13725, *"Air quality: determination of odour concentration by dynamic dilution olfactometry"*.

Grand as it sounds, olfactometry (from the Latin for "smell" and Greek for "measure") is just sniffing with style. An air sample is presented to a panel of up to ten individuals, who each take a noseful. The panellists then each identify whether they can detect the odour. If more than half of them can identify the odour, it is diluted by a known amount of odour-free air, and the process is repeated. You keep going until only half the panel members can detect the smell. At that point, you count back to see how much the original air sample has been diluted. If, say, it's been diluted by 150 volumes of clean air to 1 volume of the original sample, we would say that the odour strength is "150 odour units per cubic metre". It's important to make sure that the panellists all have a representative ability to detect odour – not too sensitive, and not too insensitive, so they give a consistent result.

That's olfactometry: quite low-tech, not very accurate, quite slow to produce results, and, I need hardly mention, expensive. You can only use it to quantify smells which are pretty strong. And, of course, it can't take account of the factors which affect an individual's response to a smell – your own personal sensitivity to odour, and all the psychological factors which might affect how you perceive and respond to an odour. But it's the best method we have: it's the gold standard for odour monitoring, if gold were transmuted into a slightly shabby, inconvenient and unsatisfactory material while retaining its stellar price tag.

Using this technique as the starting point for odour measurement means that an odour with a strength of 1 Odour Unit per cubic metre is a pretty feeble smell. It's a smell which can just be detected by 50% of the population under laboratory conditions. If you encountered 1 Odour Unit per cubic metre in your home or walking in the street, which you do all the time… you'd probably not even notice. Indeed, outdoor air typically has an odour strength of tens of odour units even when it's not noticeably smelly. These smells come from cars, gardens, fires, agriculture, cooking, pets, body odour, perfume, cigarettes, last night's beer – you name it. Odours start becoming interesting when we're dealing with odour strengths in the hundreds, thousands or (occasionally) millions of Odour Units per cubic metre. My personal record is dealing with untreated odours from animal rendering processes, which can register tens of millions of Odour Units per cubic metre. Now, that's a smell.

Alongside the gold standard of olfactometry, there are many other supporting methods for estimating and measuring odours. Some of these involve individual surveyors walking around a source of odours or a local community and recording odours against a specified list of criteria – typically strength, persistence and offensiveness. This can be useful information, but it certainly shouldn't be mistaken for a gold standard measurement of odour, or much less an indicator of odour nuisance or the experience of residents. Sometimes chemical measurements can be used to support an assessment of odour, particularly if a possible source of odours releases a characteristic set of chemicals. More recently, systems known as "electronic noses" have become available. These are instruments designed to replicate the operation of the human nose by providing a response to individual chemicals. It works through pattern recognition and learning about the kinds of odours which trigger different types of response from actual real people, and from the instrument. That makes it useful for some applications – for example, as a smoke detector – but until we know more about how our minds and bodies respond to odour, it seems to me that there will always be something a bit unsatisfactory about this approach. I don't think we fully understand the processes that it's trying to replicate.

That's why the assessment and management of smells an appropriately heady mix of science, engineering, psychology and (occasionally) guesswork. Even taking air samples for odour analysis has a dash of the absurd in it. You need to use a large rigid airtight container, for example, a blue plastic barrel with a nearly tight-fitting lid. Using this particular piece of equipment requires a specific technique to enable an airtight seal to be made, which is known to an inner circle of odour measurement technicians as "sitting on the barrel". So I once spent a reasonably embarrassing afternoon in the streets of Hull taking air samples at eight separate street corners near a cocoa factory by means of sitting on a barrel for half an hour at a time. And in all that time, nobody once asked me what on earth I was doing. Walking around with a clipboard and noting how strong the smell of the cocoa factory was would have been a lot less embarrassing, but then of course, we wouldn't have numerical gold-standard data to work with. And you don't stop looking like an idiot once you've got your samples. After completing a survey, you'll end up with twenty-five litre clear plastic bags full of air. These need to be sent to a laboratory and analysed within thirty-six hours, or the measurement isn't valid. There are plenty of courier businesses out there who will happily guarantee overnight delivery, which is great. However, not surprisingly, a bag full of air – or even eight bags full of air – don't weigh very much. You get some funny looks when you rock up to the courier's depot with three large cardboard boxes which weigh essentially nothing. "What's in them, then?" "Air." Yes, there isn't quite enough air at the destination, so I'm sending them a bit extra.

So that's how we assess odours. As well as measuring odours in the environment, we can also use olfactometric sampling and analysis to measure odour emissions from chimneys and buildings. We can even adapt the technique to measure odour emissions from surfaces – for example, sewage settling tanks, landfill sites or piles of compost. To measure surface emissions, we have to use a nifty piece of equipment called a Lindvall Hood. This is a flat rectangular box, open on one of the large flat faces. It's placed on the surface, sealed up so far as possible, and clean air is pumped through the box at a known rate. We then measure the odour level in the box. Combining the odour level in the box with the flow rate of air through the box gives us the release rate of odours from the covered area. That can be a really handy technique to enable the odour impact of the activity to be evaluated using a dispersion modelling technique. It's a tool for investigating odour problems and for forecasting odours likely to result from a similar activity which might be planned for elsewhere. Again, it's not perfect, and fraught with difficulties, but it's the best we have.

Using a Lindvall Hood on a compost heap[92]

DEALING WITH SMELLS

Information on odour release rates from chimneys and surfaces can, at least in theory, be used in a computer model to forecast the impact of the activity on odour levels in the surrounding area. This, in an ideal world, is a good way of investigating ways to deal with odours from sites which are causing a problem. The trouble is that odour measurement is not very accurate, and as we've already seen, many factors affect how odours are perceived in addition to just how strong and unpleasant the odour is.

Usually, the solution to odour problems is a combination of technical fixes, management measures and improved communications. There aren't many smell-producing activities which can't be controlled with the right approaches. Not every sewage works smells, not every farm causes widespread manure odours, not every composting site generates hundreds of complaints of odour each year. Technically, odours can usually be managed, even if it might cost a bit more than the operator was hoping. But an operator could potentially spend a lot of money to sort out an odour problem, but still experience ongoing complaints and bad publicity because of their past record. For example, an £80 million investment in a new sewage works in Reading dealt with a long-standing odour problem from the old sewage works (known as the "Whitley Whiff") when it opened in 2004. Twelve years later, when an odour affected Reading again, the first suspect in the dock was the Whitley Whiff. Was it the sewage works? Certainly guilty until proven innocent. Of course, the old sewage works was long gone, so it definitely wasn't that, but when you've got a handy alliteration up your sleeve, old habits die hard for headline writers, and the Whitley Whiff it was. The most likely cause of the 2016 reincarnation of the Whiff appears to be our old friend, agricultural slurry spreading.

£80m works could end town 'whiff'

BBC website, 23 March 2004

IS THE WHITLEY WHIFF BACK IN TOWN?

Get Reading, 4 May 2016

And it's worse if the operator of a problem site doesn't take time to meet with local communities, listen to their experiences and views, discuss the problems and reasons why odours have occurred, and outline the steps being taken to deal with the smells. It's pretty simple in concept, but it takes time and a willingness to engage with neighbours who may be angry, at a time when the operator may well not be able to give people everything they want.

DEALING WITH DUST

Sometimes you can smell air pollution, and sometimes you can see it. Airborne dust can be as big a problem as persistent odours. The photograph shows how intense dust levels can be in Kathmandu, and from my experience of spending a few days in that chaotic city, the photograph is no exaggeration.

Airborne dust on the main road into Kathmandu from the west[93]

The dust results mainly from twenty-first century traffic (well, twentieth century maybe) trying to negotiate roads constructed to mediaeval standards. The main route out of Kathmandu to the west, towards Pokhara and the Annapurna and Manaslu trekking circuits, is a dirt road through most of the city. The result: mud when it's wet, dust when it's dry. And the dust ends up everywhere: on houses, cars, shops, children, and food.

Kathmandu is by no means unique, of course: dust is a fact of life worldwide, but it's a particularly acute problem in cities without adequate road infrastructure, and where construction of buildings and roads is taking place rapidly and indiscriminately. Dust can also be whipped up from deserts or anywhere where winds are strong enough and topsoil is loose enough to make a dust storm or "haboob". Although these are natural events, they can be made worse by agricultural practices, and in particular because of the ongoing expansion of deserts worldwide.

As well as being a nuisance and a possible safety hazard due to poor visibility, dust can also affect plants by blocking up their stomata and interfering with photosynthesis. The good news is that our respiratory systems are pretty effective at removing visible particles from the air, giving us the paradox that what you can see in the atmosphere probably won't harm you. And as a little bonus, the particles that end up getting trapped in the hairs and mucus in the nose give us a useful indicator of air pollution: the infamous black snot. If you live your life in an area away from dust and particulates, when you go into the smoky city, your snot is no longer that beautiful pristine greeny-yellow colour. It turns brown or black: that's evidence that there's a lot more airborne dust and smoke in the air than you're used to, and shows that your respiratory system is doing its job. And if you live in a polluted area, the chances are that discoloured mucus is a fact of life.

So what can be done about dust? In terms of man-made sources: quite a lot. In terms of natural sources: still quite a lot, but the solutions are much more long-term and wide-ranging in terms of changes to water resource management and agricultural practices.

The most common man-made sources of dust are traffic movements, construction and demolition, storage and handling of minerals, and agriculture. Simple measures like minimising the drop heights of dusty materials and damping down stockpiles can go a long way to reducing dust impacts which result from construction sites and minerals sites. The hard part in many cases is to ensure that the most appropriate good practice measures for dust control are understood and implemented. The people who need to take steps in practice to reduce dust are the foremen and workers on a construction site. Control of dust is unlikely to be their number one priority, but it's an important part of running a construction project.

Reducing dust from traffic movements on unpaved roads also has a conceptually simple solution: surface the road. Of course, that has all kinds of cost and practical implications, but surfacing an existing dirt road is likely to be an effective investment from many perspectives: road safety, improved journey times, better access for emergency services, reduced noise – as well as less dust.

Because dust is so visible, it seems like measurement of dust is not necessarily a priority. And indeed, when we are looking at building dust controls into the design of a new development, we do focus mainly on the controls to be used, rather than on quantitative measurement and prediction of dust levels. Dust observations can and should be part of a dust management plan for large-scale construction projects.

We generally only use dust monitoring to help with dust management at sites where dust is likely to be a long-term issue, such as a quarry. Or it can sometimes be helpful at a specific problem site. I did once spend a few days in a back garden next to a builder's yard in Halesowen, trying to measure dust deposition levels to support possible court proceedings. As I recall, a combination of a leafy corner of the West Midlands and a few showers of rain meant that the results weren't very exciting, certainly not up to Kathmandu standards.

Measurement of airborne dust is often carried out by light scattering instruments, which can provide minute-by-minute concentrations of particulate matter in sizes ranging from total particulates right down to ultrafine particles with a diameter less than 0.1 microns ($PM_{0.1}$). So far, so high-tech. But when it comes to measuring the rate at which dust deposits onto surfaces such as houses, cars, food or children, our available techniques are rather more basic. The starting point is an old warhorse known as the "Frisbee gauge". Why is it called a Frisbee gauge? Well, take a Frisbee, turn it upside down, and there is your sampling technology. Almost. The design of the Enhanced British Standard BS1747 Part 1 Dust Deposition Gauge is specified in reasonable detail, but it is more or less an upside down Frisbee with a hole in the middle.

A Dust Deposition Gauge[94]

All you do is set the gauge up somewhere useful, go away for a month, return to wash any dust in the gauge down into the bottle (removing twigs, beetles etc), then evaporate off all the water and weigh the amount of dust you're left with. You can then convert this to a dust deposition rate in terms of the mass of dust deposited per square metre over the one month period. Simple, slow, not particularly accurate – but probably good enough to identify any severe dust problems. If necessary, it's possible to analyse the physical characteristics and/or chemical composition of the dust to help in identifying where it came from.

As a slightly more advanced option, directional dust flux measurement techniques are available. These use a sticky pad wrapped round a vertical cylinder to capture dust particles as they are blown past. This gives an indication of the directional variation in airborne dust levels. When this is combined with meteorological measurements, it can be a useful method of investigating and understanding the sources which are contributing to airborne dust levels – and ultimately, hopefully, dealing with them.

Almost without realising it, we often develop our own practical airborne particulate matter monitoring techniques. A clear view of the world around us is an important part of quality of life, and the presence of particles in the atmosphere can make a significant dent in visibility. If there is a particular view that you see every day, or that is particularly important to you, it's aggrieving if high levels of particulate matter (dust, smoke and aerosols) obscure the view. You might think that the particles are nothing to do with you, but they are interfering with your enjoyment of your environment. And noting how far you can see, or how often a particular view is obscured, is a way of measuring the airborne pollution. These kind of informal indicators are particularly useful to help us understand trends: is the problem getting better or worse? How does it vary with time of day, or with the season of the year? Is it better or worse at weekends or during school holidays? If enjoyment of the surrounding landscape is important to you, you'll probably be thinking about these questions without even trying. To take it further than something interesting to think and talk about, all you'd need to do is write down your observations, and you're qualitatively measuring air pollution.

Airborne dust and odour: two very tangible and practical air pollution problems that people experience every day. Dealing with these impacts, which can have a genuine effect on the quality of life for communities affected by them, requires a unique combination of practical controls, scientific analysis, imperfect data, and an understanding of people's experience and frustrations. This diverse combination of factors is what elevates a potentially unpleasant analysis of sewage, rubbish or grime into an interesting and (if we can sort out the problem) rewarding challenge.

CHAPTER 8

YOUR FAMILY, YOUR BODY, YOUR HEALTH

Within Your Own Body

One problem with dealing with air pollution is that it's often invisible, and we humans don't have any built-in way of telling when it's there. When my mum used to tell my sisters and me (or maybe it was just me) to go out into the fresh air of urban Essex in the 1970s, I'm pretty sure in retrospect that it was not very fresh at all. But we can't generally tell when air pollution is high or low, unless we can smell or see it. And the pollution which has the biggest effect on our health is invisible and odour-free. Nevertheless, we can tell when there are high levels of large dust particles in the atmosphere because we can see the haze or clouds of dust, or we can feel the particles in our noses and mouths.

High levels of dust, like those on the way into Kathmandu shown in the photograph in Chapter 7, are important in affecting quality of life. Nobody wants to live in a dusty environment, or feel the gritty crunch of dust between your teeth. But this kind of dust isn't responsible for the most serious health impacts. The fact that you can feel it in your mouth and nose shows that your respiratory system is doing its job of filtering out high levels of dust.

HOW DOES AIR POLLUTION AFFECT OUR HEALTH?

John Evelyn, who we met in Chapter 6, suggested in his *Fumifugium...* that soot in the atmosphere is responsible for some of the ills which afflict London's residents, who would otherwise enjoy "wholesome and excellent Aer".[57] In particular, the smoke (or smoak) from burning dodgy coal in the wrong places was held responsible for coughs, low birth rate and high mortality, health issues which are still linked with air pollution today. Not to mention exasperation and putrefaction of the humours, which seem to get less attention from health professionals nowadays. If only I could have a day off work every time I suffer from exasperation.

So, if it isn't exasperation or putrefaction of the humours, what happens when air pollution gets past our defences, and inside our bodies? We can sometimes detect when levels of oxidants in the atmosphere are high – for example, ozone or nitrogen dioxide. During a pollution episode, high levels of these pollutants can trigger a response in the throat and lungs, which might result in wheezing and a feeling of shortness of breath. When levels are very high, ozone and nitrogen dioxide can trigger asthma attacks in people already suffering from asthma. And we can sometimes detect chemicals in the atmosphere if they have a distinctive smell, and are present at sufficiently high levels for our noses to detect. There aren't many individual chemicals which are smelly enough for us to detect in this way, but when they do crop up (when organic waste in a drain or sewer decomposes to form hydrogen sulphide, for example), boy do we know about it.

But no one can detect the very fine particles or long-term levels of nitrogen dioxide which are responsible for so many deaths each year – thousands in the UK, millions worldwide. Whether you're in a party with a happy atmosphere, in Wales, in a polluted city, or in the downstairs toilet, one thing you'll be doing is breathing. And there's no doubt that breathing is good for you. I've been a choral singer for many years, which does involve careful attention to breathing, among other things. I once had a few singing lessons with a very patient lady who told me that I had been breathing wrongly all my life. When we breathe in deeply, the diaphragm should drop to draw air into the lungs. For some reason, I've been raising my diaphragm to take a deep breath, in a move that can only be counter-productive to deep breathing. Wikipedia nailed it home to me: *"Diaphragmatic breathing is also widely considered essential for best possible singing performance."* So I've been delivering sub-standard performances all these years, which maybe explains why my career as a vocalist hasn't made much progress. Still, I can usually provide enough oomph to sustain the long notes, and I was too set in my ways to fundamentally change my breathing habits, so I said thank you very much and decided not to pursue the singing lessons.

However we go about it, breathing is something that all of us do, every few seconds. We can't help it: breathing is an involuntary reflex that we can't normally control. You can hold your breath for short periods, of course, but sooner or later the reflex kicks in to save your life. Having said that, the world record for voluntarily holding one's breath is an incredible twenty-four minutes and three seconds – take a bow, Aleix Segura Vendrell, if you can still stand up.

When we breathe in, we take in everything that's in the atmosphere. All those inert gases, all that oxygen, and along with them, those tiny amounts of pollutants which make all the difference between good air quality and poor air quality. How do those tiny levels of pollutants cause so much damage to our health? I am no physiologist or doctor, but in the simplest terms, we invite these particles in to the most sensitive parts of our bodies. By the simple act of breathing in, we are exposing our lungs and bloodstream directly to the air and to everything in it. The small amounts of air pollutants in each breath pose a small extra challenge that everyone's body has to deal with. Your heart has to work a bit harder to push the blood round. There's a tiny,

tiny increase in your risk of contracting lung cancer. There's a small additional risk of inflammation and infection. For an individual, especially a healthy specimen, none of these little increases in risk register in our consciousness. But when we multiply up by the billions of people affected by poor air quality – and it is billions – then we see how an individually tiny increase in the risk of a heart attack might result in a few extra hospital admissions each year at a district level, or hundreds across a large city, or thousands across a country, or millions worldwide.

That's why all this breathing makes air pollution as deadly as tobacco, and worse than passive smoking, obesity, contaminated water and road accidents. And you can't cut down on your breathing like you can with cigarettes or deep-fried Mars Bars. Every few seconds, here comes another load of air with bucketfuls of life-giving oxygen, and tiny amounts of potentially harmful pollutants. We can't stop breathing, so in the long term, the only option for cutting down on the pollutants is to spend more time somewhere with cleaner air. Some people do exactly that: for many years, patients suffering from tuberculosis sought refuge in mountain sanatoriums to benefit from the clean air, for example. But for many people who might be suffering as a result of air pollution, deciding to up sticks and go somewhere with cleaner air is not a viable option, whether for financial, family or legal reasons.

While the shortening of life is sadly real, we can't usually point to individuals whose deaths were hastened by air pollution. We only know about the dramatic and huge impact of air pollution on public health because of complex statistical analysis linking exposure to $PM_{2.5}$ to hospital admissions and mortality rates. But that may be about to change, as an inquest into the death of a nine-year-old child, Ella Kissi-Debrah, gets under way. The inquest will investigate whether air pollution contributed to Ella's asthma attacks, in the light of "striking associations" between her hospital visits and episodes of high air pollution.[95]

The circumstances surrounding Ella's death are tragic and unusual. More commonly, while you might suspect that air pollution has affected your child or an elderly relative, it's very difficult to be sure. This makes it harder to communicate, explain and accept the reality of the impacts of air pollution on our health, and obtain commitments to take action. At least, it did until recently. While some people have been deeply interested in air pollution and its effects for many years, very recently, air quality has started to gain more widespread resonance among the public at large. A combination of improved communications of the science and health effects of air pollution and high levels of air pollution in many parts of the world has resulted in much greater awareness of, and interest in, air pollution. This means that the health effects of air pollutants are being taken seriously by both the public and by those responsible for evaluating and improving environmental quality, opening up real prospects for effective actions to improve air quality where levels are high. Courtroom dramas have seen the UK go back to the drawing-board to improve its plans for dealing with nitrogen dioxide in the shortest possible time. The BBC picked up on this interest, and ran a series on air quality during 2017 entitled *So I Can Breathe*.[96] In 2019, *The Times* has picked up the baton, or jumped on the bandwagon, with its campaign for *Clean Air For All*. Elsewhere, the Asian Development Bank is providing two and a half billion dollars of loans to improve air quality in China over a five year period. That doesn't just pay for a report on the air quality problem: that kind of serious money is enough to start making a real difference to air pollution.

COMMUNICATING ABOUT AIR POLLUTION

But without some kind of measurement device, no one can detect the very fine particles or long-term levels of nitrogen dioxide which are responsible for so many deaths each year – thousands in the UK, millions worldwide. I'm pretty sure that's one reason why it's taken so long for air pollution to rise up the agenda. But rise up it has, so that at last people nowadays are talking about air pollution in the same terms as we talk about obesity and smoking. It's sometimes hard to grasp the issues, because air pollution is so annoyingly invisible: we have to rely on air quality monitoring and modelling studies to tell us what the levels of these pollutants are. A cigarette or a deep-fried Mars Bar are so much more tangible causes of ill-health. We'll take a look at how we measure and forecast levels of air pollutants in Chapter 9 – but what comes out of these measurements is numbers and graphs, not something that you can physically see or touch. In 2015, journalist Chai Jing produced a documentary analysing air pollution in China (*Under the Dome: Investigating China's Smog*) in which she shows a filter paper blackened by the fine particles in the Beijing air which she breathed during just one day. If you have access to YouTube, you can see it too. That's a bit more tangible and shocking, but still, it's just one example of one pollutant from one time and one place. At the end of the day, when dealing with air pollution, we inevitably end up with strings of numbers – measured levels, standards, model results, forecasts... We often try to show these numbers using graphs and charts, but if you aren't good with numbers it can be difficult to understand what's going on.

And the same is true of some of the greatest health impacts: those seven million deaths worldwide each year. That in itself is an inconceivably large number. I can't imagine what seven million people look like: all I can do is put this statistic in the category of things in my mind labelled "a lot". I can tell you that it's pretty close to the entire population of London, or the number of breaths you take every year, or the number of steps the average British adult takes in six years, or the number of words in seven complete sets of the seven Harry Potter books. There are about fifty-five million deaths every year worldwide, so what this number says is that about one in eight of all deaths happen earlier than they would otherwise do, because of the effects of air pollution. Yes, let's deal with passive smoking, and obesity, and water pollution.

But let's also face up to the one environmental risk factor which outweighs all of these put together: air pollution.

DEALING WITH UNCERTAINTY

Dealing with air quality throws up a complex interaction of science, technology, economics, politics and psychology. It's an active area of scientific research with huge public health implications, which means that there are always new developments, and as a result, every so often changes to the established orthodoxies. There is also a lot of activity in developing new technologies which are highly relevant for air quality. But who's to say which technologies hold out genuine hope for improvements in air quality, and which are likely to have as much impact as the Sinclair C5? The what? Exactly.

And where you have that kind of potent mix, with scientists, some parts of the business community, environmental specialists and increasingly vocal members of the public trying to tell politicians and businesses that they need to do something, and what they need to do, you can end up with some questionable decisions.

Sometimes, there is a lot of pressure to take rapid decisions based on inadequate information. Both the decision itself and its timing can be driven by a wide range of factors, not just from a desire to identify and implement the best possible outcome. Politicians and regulators may need to take early action before robust information is available, or may need to be seen to be acting decisively, for example. The 2001 outbreak of foot and mouth disease among livestock in the UK, for example, is estimated to have resulted in an overall cost to the UK economy of nine billion pounds.[97] That might well have been money well spent – but it is a lot of cash, reflecting a cost to rural tourism of about three billion pounds, as well as the costs of managing the disease itself. Could alternative approaches have been adopted which delivered as good an outcome, or nearly as good an outcome, or maybe even a better outcome, but at lower cost? Maybe they could, but I mention this as an example of a situation where decisions were taken on managing the disease based on a complex mix of scientific, veterinary, environmental, political and social factors. Throw a bucketful of uncertainty into that mix, and you might even conclude that bringing it in under ten billion wasn't a bad result.

How does that relate to air pollution? Well, we don't (yet) have the same sense of urgency that surrounds the outbreak of a disease like foot and mouth, or BSE, or Zika virus, or Ebola. Perhaps we should: anything which causes seven million early deaths each year should surely have some urgency about it. We're not in a position of having to take urgent decisions with incomplete information, as those responsible for managing epidemics might well be. But there is a pressing need for action, and maybe some disagreement about the best ways to secure change. What are the main contributors to air pollution? How much would it cost to reduce or eliminate these sources? Should we focus on technological solutions or on societal and behavioural change? Should we place restrictions on what individuals or businesses can do – such as limiting car use to odd and even number plates on alternate days, or closing down industries during air pollution episodes? Should we pour more resources into enforcement of existing legislation, or are new controls needed? How do we balance action and investment between countries such as the UK where pollution levels are relatively low but still resulting in tens of thousands of deaths each year, and countries such as India where air pollution causes over a million premature deaths annually? How much should we invest in improving air quality, and who should pay? So many questions, and often no complete answers.

I've also found that, surprisingly often, the world of air quality science and policy is conviction-driven. Many people, probably including myself, hold views which are driven by conviction rather than by hard evidence or scientific logic. Until recently, it was commonly held that diesel cars were a better environmental option than petrol cars, because of the better fuel efficiency. That makes diesel cars preferable from the perspective of climate change. Diesel engines also typically last longer than petrol engines, so that's another point in their favour. But then a few years ago, we woke up to the impacts on our health of particulate matter and NO_x emissions. Diesel cars, particularly older diesels, emit more of both pollutants than petrol cars, and (as we've already established) air pollution is harming people today, whereas the full impact of greenhouse gases on the global climate remains some way off. So nowadays, it seems more environmentally responsible in the majority of cases to drive a petrol car, or better still a hybrid or electric car, or no car at all if that's an option. In the UK, the tax system is slowly catching up, with the annual road tax more or less equal between petrol and diesel cars from 2017 onwards. Looking at car choice simply as a binary decision between petrol and diesel, it's an area where received wisdom and convictions have changed over the past few years, away from diesel and back towards petrol. And the tide is now turning in a different direction (if a tide can have more than two directions), as Volvo has just announced its intention to become the first major manufacturer to fit electric engines in all its cars, either as hybrids or fully electric vehicles. Every new model of Volvo car from 2019 onwards will have an electric engine, either as a pure electric vehicle or a hybrid. This ties in with commitments from the UK and French governments to ban new petrol and diesel cars from 2040. The roads will look and sound very different by the middle of this century.

Without doubt, that's a step in the right direction for air quality, because a car operating on electricity does not directly emit pollutants at the point of use. Hybrid cars work in a different way, but do still emit lower levels of pollutants than conventional petrol and diesel cars, for example, by avoiding the need to run a large engine to provide the poke needed for acceleration and hill climbing. Electric vehicles have no direct emissions: instead, the sources from which air pollutants are released are moved away from the immediate vicinity of people living near to the roads to the point of generation, where they can, we would hope, be better controlled. And there are many low or zero emissions technologies for generating electricity which help to reduce the overall air pollution impact of electric vehicles. Electric and hybrid technologies are not the last word in avoiding emissions from road traffic, because all that electricity has to come from somewhere, and the non-exhaust emissions (fine particles from brakes and tyres) would also continue. So-called "well to wheel" studies seek to understand the overall environmental impact of different vehicle power technologies. These studies compare the quantity of pollutants emitted per kilometre travelled, taking account of all stages in the fuel and energy production processes. There are a lot of variables to consider: the type of car, the type of engine technologies, how and where the car is driven, and the mix of technologies used to generate electricity to name but a few. And that's before you start to think about whether the pollutants are emitted from the vehicles themselves, or from a power station somewhere else, and how that might affect the impact of the air pollutants. But to draw a sweeping generalisation, moving from conventional to hybrid and electric vehicles will deliver improvements in urban air quality by reducing or eliminating exhaust pipe emissions from individual vehicles, even though there will be an increased need to generate electricity. It's progress.

And it's not only cars where opinions diverge, although they do give the boys from *Top Gear* plenty to talk about, with even the occasional throwaway mention of exhaust emissions. Jeremy Clarkson, former presenter of the estimable motoring show, has spoken out in favour of hydrogen fuel cells, citing the large quantities of hydrogen in the universe and the low emissions. What's slightly unclear is how the hydrogen in, for example, the Horsehead Nebula gets from there, 1,500 light years away, into the engine of your nippy little Hyundai ix35. Even coming back, more rationally, to planet Earth, there are difficult questions to be answered about the cost of producing and transporting hydrogen. Where hydrogen cells win over battery powered electric vehicles is in the speed of refuelling: it's a few minutes to refill a hydrogen tank, compared to several hours charge time for an electric car, although recharge times are coming down to below an hour for some electric systems.

DEALING WITH AIR POLLUTION FROM INCINERATORS

There are other important questions where convictions remain as entrenched as ever. I have spent a lot of the past fifteen years investigating the air quality and health effects of waste incinerators. I can't think of any topic which has generated more controversy and more ill-informed opinion over that time, including the question of whether we should run our cars on interstellar gases. I don't just mean in the small world of air quality – I mean any topic at all. Indeed, when I first began to look into the air quality issues, and particularly the health issues, which surround waste incineration, I could hardly believe some of the claims being made, both in favour of and against waste incineration.

For example, I have stood next to representatives of the waste industry who have claimed that waste incinerators actually clean up the air – the claim was that the pollution levels coming out of an incinerator chimney are actually lower than the levels of air pollution in the air. Well, for the avoidance of doubt, this is absolutely not the case. If it was true, we wouldn't need to worry about all those tedious and expensive chimneys. No, nitrogen dioxide, for example, is emitted typically at levels about a thousand times higher than those in the atmosphere. Or maybe ten thousand times higher. That's why we need chimneys, to make sure that waste to energy plants don't affect air quality for people living and working nearby.

On the other side of the equation, I once went to a public meeting and listened to my local Member of Parliament explaining that a new incinerator planned for our town would result in increased cancer and infant deaths. And that, as I'll go on to explain, is equally misleading. To be honest, I can't really blame him for what he said – he was only repeating claims that so-called expert advisers had told him, and any politician can spot a bandwagon to jump on. He later said that he would "lie down in the road" to stop the incinerator being built. What seems harder to understand was his viewpoint after planning permission had been granted for the incinerator, that "it is now time to accept that the project will go ahead". Yes, it was time to accept what he had previously asserted was a cancer-causing, baby-harming incinerator… I can only hope that, by that time, he no longer believed the claims he had been fed by people who, for whatever reasons, didn't want the incinerator to go ahead.

This is a subject which has generated so much controversy, anger, information, misinformation, and hot air that it's worth looking at in a bit of detail. Quite a lot of detail, in fact. I think the question of whether incinerators affect our health is fascinating and important in its own right, and what it tells us about the interaction of environmental science and public perception is also highly interesting, not to say topical, as we in the UK launch ourselves into the uncharted waters of hydraulic fracturing.

So let's start from the top. Firstly, I'm not going to distinguish between the terms waste incinerator, waste to energy plant, and energy from waste plant. All household waste incinerators in Europe now have to recover energy, so there is no such thing as a straight incinerator which just burns rubbish. The industry says "waste to energy", and people living nearby say "incinerator", so I tend to use the terms interchangeably.

Next, I'm not going to go into the question of whether waste to energy plants are a good thing or not in principle. Decisions about waste facilities are based on a complex mix of factors, uppermost of which is that incineration is near the bottom of the UK's "waste hierarchy". That means that all options for minimising waste production and maximising recycling have to be taken before consideration is given to energy recovery. If a waste incineration facility is needed, the waste hierarchy will also help to inform how large such a plant should be, and what waste streams should be used for disposal. That's a wide-ranging decision which needs careful consideration of trends in waste and recycling, and a well designed and implemented waste strategy. Every waste to energy plant I've been involved in has faced intense scrutiny about whether energy recovery is the most appropriate technology for dealing with residual (non-recyclable) waste, and if it is, how big should the plant be to accommodate future changes and trends in waste arisings. I won't go into these questions here, because the first association in many people's minds when they hear the word "incinerator" is not whether it's the right size, or what it should be called. It's the question of health impacts on local communities. For many people, this is a visceral, deeply felt response, even for individuals who don't have any particular experience or knowledge of what waste to energy plants are. The deep-seated fear that some people feel is real, and I am sure it can have its own effects on health – sleepless nights, stress, and feelings of nausea for example. But is it reasonable to have this almost instinctive response to waste incineration? And why do so many people feel this way?

SHOULD WE BE WORRIED?

To answer the question, "is it reasonable to be fearful of the health effects of waste incineration?" my starting point is a study that I and my colleagues carried out for the UK government Department for Environment, Food and Rural Affairs back in 2004. The aim of the study was to look at all different kinds of waste management infrastructure, and carry out an independent assessment of the evidence for health and environmental effects. Looking back, the report that we produced brought together a lot of useful information into one place, enabling us both to compare different waste and resource management options, and to set the environmental and health consequences in some kind of context with other risks that we live with every day. And I think we were able to draw some useful conclusions. We quantified environmental emissions to air, land and water from different types of waste facilities. We highlighted the evidence for environmental issues arising from waste facilities – landfill and composting odours, for example, and the risks of groundwater pollution from landfill sites. We evaluated the evidence for adverse health impacts of landfill sites (slight, but non-zero), and also evaluated the many reports looking at the health impacts of waste to energy facilities. And there is a lot of evidence out there. One of the features that we found is that, with waste to energy facilities, you know what you're getting. The operations are monitored continuously, and the emitted substances are discharged through a chimney, so of all the waste activities that we considered – indeed, of almost any industrial activity at all – waste incineration is the best understood.

When you put all the evidence on health together, a pattern begins to emerge, and it's a pattern which has not changed in the years since our report (*Review of Environmental and Health Effects of Waste Management: Municipal Solid Waste and Similar Wastes*) was published. The non-headline grabbing message is that waste incineration carried out to current standards has no detectable effects on health. That doesn't mean that there is no effect at all, but if anything is going on, it is too small to detect using our currently available techniques. Putting it another way, even if you live very close to a waste to energy plant, all sorts of other factors are way more important for your health than what's going on at the incinerator. These complex factors include lifestyle (how much exercise an individual takes, for example); hereditary factors such as susceptibility to particular diseases; diet; other existing health conditions (such as asthma, bronchitis or obesity) and, of course, smoking. From my reading of the public health literature, anyone who's a dedicated smoker really doesn't need to worry about any environmental factors which could affect their health. The effect of smoking on health outweighs everything else by miles, so it'll almost certainly get you before anything environmental does.

All these complex and highly individual factors which affect health make it hard to distinguish the influence of individual environmental matters. Hard but not impossible: the science of epidemiology and its detailed statistical techniques are designed specifically to tease out the contribution of different causes to ill-health. These techniques were used to highlight the substantial contribution of smoking to the incidence of lung cancer during the first half of the twentieth century, and to carry out the much harder analysis to identify the contribution of the London smogs to mortality in the 1950s. More recently, epidemiological analysis has been applied to quantify the overall contribution of $PM_{2.5}$ and nitrogen dioxide to mortality rates and the incidence of respiratory disease. Yet when we try to use the most advanced techniques to tease out the effect of living near an incinerator on public health, they are not powerful enough to find any detectable impacts. The latest national UK study published by the Small Area Health Statistics Unit at Imperial College couldn't find any evidence of increased health risks for infants related either to incinerator emissions, or living near an incinerator operating to the current regulations.[98] All of which is reassuring, even if it doesn't make for attention-grabbing headlines.

THAT'S A BIG "IF"...

You might have noticed that there is a reasonably big "if" written in to this generally reassuring conclusion. The conclusion is applicable to waste incineration *if* it's carried out to current standards – by which I mean, waste incineration carried out in properly designed, located, operated and maintained facilities. What about incineration in times gone by then, when standards were not as tight as they are now? And what about when things go wrong, and a current operation is no longer complying with the current standards? These are reasonable questions, and here the evidence starts to look a bit different. Looking back at incineration in the 1950s and 1960s in the UK, there is some evidence that the plants operating at the time could have resulted in a small but detectable increase in the incidence of cancer. The plants operating at that time bore no resemblance to the waste to energy plants running today in the UK and throughout Europe, in terms of design, operation or monitoring. There was no regulation requiring operators to control emissions to the atmosphere from the process, and so it might be expected that emissions then were hundreds of times higher than they are today. Once waste incineration processes in the UK started being regulated in the 1980s and 1990s, then observable health effects more or less dried up. And even then, emissions of the most harmful pollutants were way higher than they are now.

Waste to energy facilities have had to conform with a series of tightening environmental standards since the 1980s, with some of the key controls being implemented under the European Waste Incineration Directive of 2000. As well as setting emission limits, this directive also set operational requirements including a requirement for flue gases to be held above a minimum temperature of 850 degrees centigrade for at least two seconds, along with operational and monitoring requirements. Putting together emission limits, operational requirements, monitoring and reporting is a winning combination which is responsible for the dramatic change in waste incineration from the 1980s to the present day.

Looking more widely, there are some plausible reports of adverse health impacts from plants scattered around the world. I've been very interested to read these reports carefully to see if they show a different picture to the good news story about current operations in Europe. And do you know what, the picture is more or less the same. For example, a study of municipal solid waste incinerators in Japan published by Tango and co-workers in 2004[99] focused on seventy-two incinerators which had particularly high emissions of dioxins and furans – these are two groups of similar organic chemicals which contain chlorine and oxygen, and are about the most toxic substances known to science. This study found some evidence that the incidence of infant deaths was higher for families living between one and two kilometres from the incinerators, and declined for families living further away from the incinerators. In this instance, "high" emissions of dioxins and furans meant incinerators emitting more than eighty nanograms per cubic metre of dioxins and furans. That quantity of anything – eighty nanograms per cubic metre – is almost unimaginably small, but just because it's small doesn't mean it's not a lot, nor does it mean that it isn't a health hazard. And indeed, the Tango study and a lot of other evidence confirms that dioxins and furans at this level are a health hazard, which is why waste to energy plants in Europe have all been limited to less than 0.1 nanograms per cubic metre since 2005 at the latest. That really is a tiny concentration. We saw earlier that the amount of particulate matter typically present would amount to the weight of a small paper clip in the air inside the Albert Hall. Well, the concentration of dioxins and furans in incinerator emissions is the weight of a small paper clip in one hundred thousand Albert Halls. It's a tiny amount, but dioxins and furans are so toxic that we need to consider even such tiny concentrations very carefully. Most plants in Europe nowadays emit less than ten per cent of this limit. So the study by Tango and colleagues is highly relevant, but it refers to facilities emitting maybe eight thousand times more of the most hazardous chemicals compared to waste incineration facilities in the UK and throughout Europe. The fact that this study did find a health effect from facilities emitting eight thousand times more than the comparable facilities in the UK shows that when there is a big enough thing to find, epidemiological analysis will find it.

It's findings like this that give me confidence that we're on the right track in concluding that it's really, really important that waste incineration facilities are properly designed, located, operated and maintained. If these standards drop, then there is a small but possibly detectable risk of an increase in health impacts

– although the evidence of studies like the Japanese analysis is that things would need to go really badly wrong before there was any detectable impact on health. Conversely, if these well-established and understood standards are maintained, then there is no detectable risk of health impacts. Nowadays, with almost twenty years' experience in operating waste to energy facilities in accordance with the requirements of the Waste Incineration Directive, and an effective system of regulation in place, things don't go really badly wrong. That's not to say they never will go badly wrong: operating a waste to energy plant is difficult and challenging, and requires expertise and ongoing investment. But I think we have the knowledge and expertise to run waste incinerators safely, and the regulatory and monitoring systems in place to pick up any problems before things go seriously wrong.

DIOXINS AND FURANS

If that's the case, then why do people continue to object to waste to energy plants on the grounds of health? One answer to that is – many people don't any more. Friends of the Earth, for example, is no friend to incineration. Its website[100] lists six reasons why Friends of the Earth is opposed to waste incineration, but the risks to health is not one of them. The UK Without Incineration Network[101] states that it opposes the incineration of waste because it *"depresses recycling ..., destroys valuable resources, releases greenhouse gasses, and is a waste of money."* No mention of health impacts here, although UKWIN does highlight *"legitimate air pollution concerns"* and the associated health risks when discussing their position in more detail. Even Greenpeace, probably the most ideologically-driven of environmental groups active in this area, seems to have let its opposition to incineration on the grounds of health concerns lapse. *"The Greenpeace website does still include the claim that "Incinerators ... create the largest source of dioxins, which is one of the most toxic chemicals known to science"* – but this is on an area of the site headlined as having been archived. The claim is still there, however, but certainly in a UK context, any suggestion that incinerators are a large source of dioxins is not true, and hasn't been true since the European waste incineration directives started to take hold in the 1990s. Take a look at the graph below:

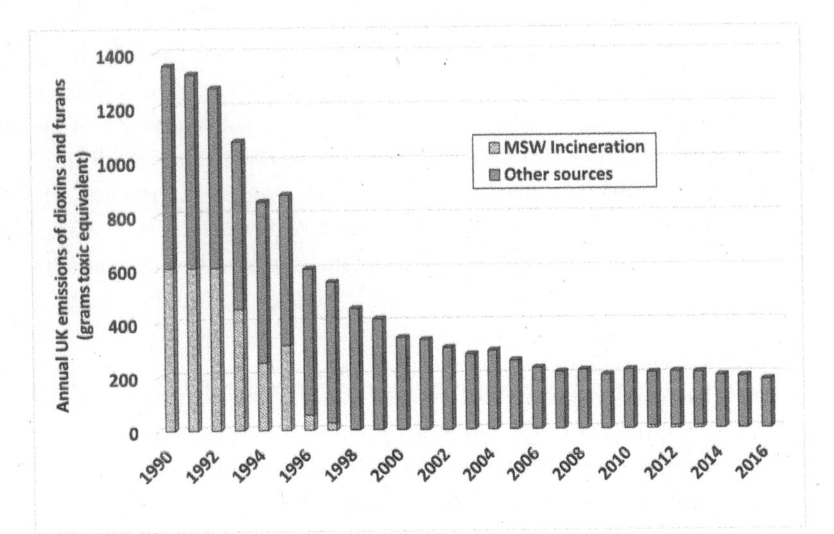

Annual UK emissions of dioxins and furans to the atmosphere[102]

This gives us a clue as to why incineration has such a bad reputation. The group of chemicals known as "dioxins" first came to public attention in a big way in 1976, when an accident at an industrial plant in the Italian town of Seveso resulted in a release of about a kilogram of one of the most toxic members of the dioxin group, 2,3,7,8-tetrachloro dibenzo para-dioxin. This chemical was known to be extremely hazardous to health and large numbers of wild and domestic animals were found dead as a result of this release. Follow-up studies were inconclusive regarding serious health effects for the local population, but there was no doubt that it was an extremely serious and potentially damaging incident. Following the Seveso accident, a system of industrial safety regulation was put in place throughout Europe. Public awareness of dioxins, and the closely related group of chemicals known as furans, was greatly increased as a result of this incident, and it became apparent that incomplete combustion of fuels containing even small amounts of chlorine could give rise to significant emissions of dioxins and furans. The figure above shows that in 1990, waste incineration accounted for almost half of the UK's emissions of dioxin and furan emissions. The finger was rightly pointed at waste incineration as the single largest UK source of dioxin and furan releases into the atmosphere. Something needed to be done.

At the same time, the European Commission was introducing controls on waste incineration for the first time. The European Municipal Waste Incineration Directives of 1989 did not explicitly set limits on emissions of dioxins and furans (although member states were permitted to do so themselves), but did set technical requirements on waste incinerators which would ensure that the basic steps for minimising emissions of dioxins and furans were in place. For existing plants, these requirements had to be achieved by the end of 1995. That's why there was a steep decline in emissions of dioxins and furans from waste incineration between 1992 and 1995, and then a sudden drop in 1996. By 1996, emissions from waste incineration were less than a tenth of the emissions in 1990. And it didn't stop there: it only took a couple of years for a further tenfold decrease, and since 2003, emissions from waste incineration have been a further factor of ten lower. That's about one thousandth of the quantity emitted in 1990. Show me any other industrial process which has reduced its emissions by 99.9% in less than fifteen years: I don't think there are any. Over that time, while incinerator emissions have reduced by a remarkable 99.9%, emissions of dioxins and furans from all other sources have gone down by a worthwhile, but less impressive 70%. By 2016, incineration of household waste accounted for 0.4% of the UK's emissions of dioxins and furans.

Now, I'm not a big fan of dioxins and furans – they are a horrible group of chemicals which stick around in the body for years, and Greenpeace is right to say that dioxins are a group of chemicals including some of the most toxic known to science. As we can see from the graph, waste incineration used to be the largest source of dioxins in the UK. But, in one of the most remarkable technical fixes ever, waste to energy plants now make a minuscule contribution to UK emissions of dioxins and furans. They haven't been the largest source of dioxins for over twenty years. So we don't need to worry about dioxins from incinerators. Should we carry on worrying about dioxins and furans from other sources? Possibly we should: a risk assessment and cost-benefit analysis would tell us whether continuing to take steps to reduce dioxin emissions to the atmosphere would be a cost-effective way of continuing to improve public health. In fact, dioxins and furans are only partly an air pollution problem. The atmosphere is the pathway through which quite a lot of dioxins move on their way through the environment. But only about 2% of our exposure to dioxins and furans comes directly from what's contained in the air that we breathe in. Exposure through the food chain is much more important. This is a multi-stage process, with dioxins in the atmosphere eventually settling out on the ground. The chemicals may land on fields where food is grown for human consumption, or may land on pastureland or crops being grown for animal feed. Dioxins might also end up in water courses, and from there they could be used for irrigation or water supplies for people or farm animals. The end result is that dioxins, which have a very long lifetime in the environment and in the bodies of animals as well as people, can end up in the food chain. Although we often think of dioxins and furans as an air pollution problem, the most recent incidents of dioxin and furan contamination don't have anything to do with atmospheric pathways at all. For example, in 2010, dioxins got into some egg products because of a mix-up at a factory in Germany, in which oil intended for biofuel manufacture was mistakenly used to make animal feed. In the most infamous incident in the UK, ash from a waste incinerator in Newcastle was used to make paths in allotments, where people were growing vegetables and keeping chickens. Again, not directly an air pollution problem, but certainly there was an indirect link because the ash came from systems used to take dioxins and other contaminants out of the incinerator flue gases, from which they would otherwise have been released into the atmosphere. The use of incinerator ash in this way was, of course, a monumental mistake: if you wanted to think of the worst possible thing to do with a material containing dioxins and furans, it would be hard to come up with a more stupid idea than putting it on the ground in an allotment. Whatever the risks to health and the environment, once you've spread the ash on the ground in a public place, it's out of your control. Which is not to say that there have been long-term health problems resulting from the use of incinerator fly ash at the Newcastle allotments: so far as we can tell, there won't be any. The lesson to be learnt from this sorry incident is the importance of maintaining proper control of waste streams. Alternative uses of incinerator ash include road construction aggregate and manufacture of breeze blocks. In both these cases, any dioxins or other contaminants present in the ash are kept in a solid matrix, and don't come into direct contact with vegetables, chickens or frolicking children. That's the way it should be.

Dioxins and furans are highly potent carcinogens, and they also have a long lifetime in the human body, of about seven years. So although we are exposed to unbelievably small amounts, we have to take the health risks posed by dioxins and furans very seriously. But one thing's for sure: there's virtually nothing to be gained from further controls on dioxin and furans from waste incineration. These days, the largest single source of dioxins and furans is burning of solid fuels in homes – including coal, coke and wood burning – which accounts for about a third of UK emissions. Combustion in industry makes a similar contribution. Open burning of waste (for example, on bonfires) accounts for about 10%. That's the place to start – indeed, open burning of waste is probably the area where a reasonably significant improvement could be delivered without any major practical difficulties. It's just a matter of changing behaviours, so that nobody burns waste materials in the open. There's no reason why waste materials have to be burnt on open fires, they just need to be dealt with through your normal recycling and waste collection services. If we could make burning waste on bonfires as unthinkable as putting your children in a car without seatbelts, we'd deal with a tenth of dioxin and furan emissions at a stroke. That's low cost, low tech, and thirty times the benefit you'd get from immediately shutting down all of the UK's incinerators (and by the way, what would we do with all that non-recyclable rubbish?).

Looking back to 1990 and earlier, waste incinerators were a significant source of dioxins and furans. While the waste industry has done an extremely effective job in delivering a 99.9% reduction in emissions alongside a significant expansion in incineration capacity, you could look at the timing of this reduction alongside the introduction of European directives, and conclude that the industry didn't do anything that it wasn't forced to do through legislation. The process improvements were introduced in response to legislative requirements, not because the industry recognised that there was a problem that needed to be addressed. Even after the Municipal Waste Directive was published in 1989, there was no noticeable progress in reducing emissions from waste incineration for several years, until the December 1995 deadline came onto the horizon, and emissions finally started to go down in 1993. To be fair, during the 1980s and early 1990s, the waste industry was not awash with money to invest in technologies which improve the environment but don't do anything to improve capacity or throughput, so perhaps it wasn't until there was a legislative stick that the necessary investments could be justified. Whatever the truth, this gives the impression of an industry which was not in any hurry to deliver environmental improvements. Together with occasional incidents of unsatisfactory operation, maybe that's one reason why the industry continues to experience a low level of trust with the general public to this day. But even if that's the reason why such dramatic improvements were made, in the end, the industry has delivered on its obligations, big time. I'm still confident that we don't need to worry about dioxins and furans from waste incineration.

DEALING WITH THE HEALTH IMPACTS

So that's the overall picture, or my take on it at least. If you happen to live near an operational or proposed waste incinerator plant, you're likely to be most interested in the effect of that individual plant on your own health, and that of your family, rather than sweeping statements about incinerators in general. So when we are evaluating individual waste to energy plants, we have to make sure that each one of them has an acceptable air quality and health impact. We have the tools available to allow us to do this to a high degree of confidence. One reason for this is that waste to energy facilities are very well-controlled and understood – as well or better than almost any other industrial process. We know what pollutants are emitted, and we have good information on how much is emitted. Another reason for confidence in the findings of air quality and health risk studies is that air quality models work best for tall, single point sources which are not strongly influenced by nearby buildings. This is often the situation for waste to energy plants, which tend to have tall chimneys (though not always in the same 100 metres plus league as the Cornwall plant).

And life is made a lot easier for the air quality assessor because of the limits on pollution emissions specified in the current incarnation of the waste incineration directive (the limits are now in Annex VI of the Industrial Emissions Directive, 2010/75/EU, in case you're interested). This means that emissions of substances like dioxins and furans and metals are at such a low level that significant impacts on air quality would hardly ever be forecast. In fact, I only know of one recent example where the air quality impact of a proposed waste to energy plant was identified as being too great. This was a proposal for a large waste to energy plant in Essex, which would treat up to 600,000 tonnes of waste per year, and under the original proposals would have had a stack height of thirty-five metres. This was considered by the Environment Agency to be too low – their view was that the air quality impact could have been significantly reduced with a higher chimney. Even with the lower chimney height, it wasn't as if air quality standards were going to be exceeded, it was just that the developer could have done better. At first sight, I would agree with this: every waste to energy facility that I've been involved in has had a demonstrably insignificant effect on air quality. And when we follow this up to look at the health risks of exposure to air pollutants from waste to energy facilities, not surprisingly we get the same answer: no significant effect on health. The air quality impacts and health effects are not zero: there's always a small addition to the baseline levels of air pollutants. But we can always demonstrate that these increments aren't significant when judged against air quality standards and other independent criteria. That Essex plant, by the way, was eventually granted a permit with a rather taller fifty-eight metre stack.[103]

In some cases, because impacts on health have been so effectively dealt with, it's the effect of emissions to air on sensitive habitats which places the greatest constraint on a new waste to energy facility. This results in higher release points (as was the case with the Cornwall facility), or lower emissions than would otherwise be needed. The key pollutant which has the potential to affect sensitive habitat sites is oxides of nitrogen, and here there is some scope to reduce emissions below the limits set in the European directive, if needed to mitigate impacts on ecosystems. Any changes we make to mitigate these impacts on natural ecosystems also have a further small benefit on air quality and health impacts for local communities.

Working on waste to energy projects through the planning system has given me the opportunity to discuss complex questions about the health and air quality impacts with local communities and decision-makers. People living close to a proposed waste to energy plant are often very worried about the effects on the local environment, and on health for themselves and their families and friends. These worries are often fuelled by rumours and half-truths (or downright untruths) about the proposals and their possible effects, and can spill over into anger. We are often predisposed to believe the worst about something new and unwelcome, which a proposed waste facility almost always is. People with an interest in opposing a new waste facility are quick to pick up on this and provide information on the possible health consequences which sounds plausible, and plays to deep-seated fears. On the other side of the argument are often people like me, who have carried out the technical analyses to make sure that there won't be any detectable or significant effects on health. But our words can sound very hollow, as we are of course paid by the developer to carry out the assessments.

People sometimes ask me whether I have ever carried out an assessment which showed an air quality problem with a new development. The criticism (implied or stated) is that we are hired guns who will just say what we're paid to say irrespective of the evidence. The answer is yes – sometimes we do show that there would be an air quality problem. However, under those circumstances, you don't just publish a report which concludes that there is a problem – this would effectively be a suicide note for the project, however welcome that might be in some quarters. Instead, you go back to the drawing-board, identify ways of dealing with the problem, make some changes and try again until you find the scheme design which works from an air quality perspective. For one such proposal, we tried numerous combinations of NO_x, sulphur dioxide and ammonia emissions before we found a workable scheme which would avoid impacts on the nearby protected moorland and ancient woodlands. Not surprisingly, the scheme which avoided impacts was the one which went forward for approval, not the other fifty which didn't. And on the question of trust: I have never put forward information that I believe to be false, and I don't think any of the professional air quality specialists who I've worked with over the years would do so either. The air quality specialist's job is often to find a workable proposal which meets the relevant air quality standards and guidelines, but there has never been any falsification or bending of the results to get the right answer that I've seen.

THE SIGNIFICANCE OF "SIGNIFICANCE"

One reason our words can sound hollow is that little word "significant". There is no getting away from the fact that a new waste incinerator is always bad news for local air quality. There's always an increase in air pollution, and an increase in health risks, however small that increase might be. So as the air quality specialist, I am always the bringer of bad news to local communities (and often the provider of bad news to the developer as well, when I tell them that they'll have to spend more money on their pollution control). I'm very rarely in a position of showing an improvement in air quality. The best I can hope for is to be able to tell people confidently that there will be no "significant" effects on health. A neighbour to an unwelcome new development is very likely to retort that they don't want *any* effects on their health, however insignificant. I completely understand that point of view, but a moment's thought tells you that it isn't how the world works. Everything we do affects other people. If you decide to drive your car, you're affecting other people through the noise and pollution your car will produce. And you're adding a bit to traffic congestion, and increasing the risks to nearby pedestrians. Those risks and impacts are tiny, but they aren't zero. We all accept that small and insignificant risks resulting from what other people do are a part of life.

Something as big and complex as a waste to energy plant can often have significant effects on other people – for example, through noise, pollution and health risks. In this case, the risks and impacts could be much greater than those everyday risks, so we have systems of deciding whether such a development should go ahead. The planning and permitting systems provide the means for a democratic society to weigh up the risks and impacts of the new development against the benefits that it will provide – for example, in meeting a need for disposal of residual waste which can't be recycled. In this context, coming up with a proposal which demonstrably has no significant impact on air quality and health makes a lot of sense. What we're saying is that when you are weighing the benefits of a proposal against its impacts through the land-use planning process, then the air quality and health impacts should not add any weight to the "impact" side of this balance. We certainly shouldn't take big decisions on the basis of avoiding risks and impacts which are demonstrably insignificant: that's a sure recipe for terrible decision-making. You would end up missing out on a lot of benefits, without securing any environmental benefits. Decisions about waste incinerators and any other big investment project should concentrate on what's significant, not on what's insignificant.

At a smaller scale, of course, we make these kind of judgements all the time. If you're in a supermarket deciding whether to buy "I Can't Believe It's So Good" (the spread that I can't believe was formerly known as "I Can't Believe It's Not Butter") or the supermarket own-brand alternative, you might decide that the twenty pence price difference is not significant, and go for the branded option. Or you might decide that the difference in flavour is not significant, and save yourself the twenty pence by buying the own-brand margarine. Either way, you've decided that an aspect of the purchase is not significant, or is less significant, and taken a decision accordingly. So, with a planning decision, it would be perverse for a planning authority to give a lot of weight to an aspect of a development which is not significant. If the air quality and health impacts of a development are not significant, then whatever planning decision is taken, it won't make any difference to the air quality experienced by local residents, or to their resulting health expectations.

So there's no logical basis for giving any weight to an impact or risk which can be definitively described as "insignificant". And indeed, planning decisions are not often made on this basis, even though many people's instinct is to reject any increase in risk, however small, as unacceptable. When air quality impacts affect planning decisions, this is more often influenced by whether the evidence on air quality and health impacts is considered to be reliable. It's very rare for there to be any substantive challenge to the air quality and health risk assessments that we carry out for waste to energy facilities. The assessment techniques are well established, and the waste incineration process and associated emissions are well understood in a lot of detail. And the emissions are now at such a low level that the air quality and health effects are demonstrably minuscule. Consequently, the main challenges to air quality and health risk assessments for waste to energy facilities focus on alternative views on the health risks of incinerators more generally. As I've already said, my view is that there is the potential for problems to occur, but as long as waste incinerators are designed, located, operated and maintained to current standards, there are no detectable risks to health. Not surprisingly, this is exactly the answer we'd expect if the air quality and health risk studies for individual waste to energy plants are giving the right answers. However, it's only fair to record that others do hold different views.

UNCERTAINTIES ABOUT UNCERTAINTY

There is perhaps a remaining question around uncertainty. Air quality and health risk studies rely on the results of models, and the models rely on assumptions about what will be emitted from the incinerators – so the conclusions of these studies must be uncertain, right? Yes, that's right. The statistician George Box famously said *"All models are wrong but some are useful."*[104] I would put properly conducted air quality studies for waste to energy facilities firmly in the "wrong but useful" category. The models are built on a sound theoretical basis, with verification using experimental data. The information about the source of pollution that goes into the models is reliable – we have an excellent database of information on emissions from this kind of plant, backed up by regulations which set limits on emissions. So we often run the models on the assumption that the process will operate at the permitted limits, and back this up with measured data from comparable operational processes to demonstrate that the limits can be achieved in practice. Also, while the models are inevitably inaccurate, we know how inaccurate they are likely to be. That's particularly useful, because we can account for this uncertainty by erring on the side of caution when setting up the model. So we can be pretty confident that the model results are overestimates of the levels that will arise in practice. Yes there is uncertainty, but there is less uncertainty than almost any other kind of air quality study, and we can account for this uncertainty when setting up the air quality model and interpreting the study results. Nevertheless, if someone is determined to find a reason to object to a new development, any uncertainty can be seized on to muddy the clear waters of a tidy air quality impact assessment.

TOUGH DECISIONS

I do feel for the local politicians who have to deal with complex technical and scientific issues when deciding whether to grant permission for a new installation like a waste to energy plant. Suddenly, people without any particular scientific background are being asked to decide whether a dioxin intake of 0.018 picograms per kilogram body weight per day is OK or not. Sometimes, you can see the members of a planning committee thinking hard about air pollution and health impacts, in the face of conflicting opinions from local residents, the applicant, the applicant, their advisors, and their own gut feelings. When councillors grapple with the issues and reach a balanced and robust view, that's very satisfying – even if it isn't necessarily a view that I would have taken. On the other hand... I particularly remember spending two full days at a council committee meeting discussing a single application for a waste incinerator, with detailed questioning of a wide range of issues in the presence of 200 members of the public. Democracy in action, I thought. At the end of the second day, after exhaustive and careful discussion of all aspects of the proposed plant, the vote was taken as to whether planning permission would be granted. The committee members all voted exactly on party lines. We needn't have bothered.

AIR QUALITY AND HEALTH

So we've covered a lot of ground while looking at what goes on within our own bodies, as air pollutants worm their way deep into our poor vulnerable lungs and onwards to put a small extra strain on our respiratory and cardiovascular systems. These effects of air pollution on our health are hard or impossible to discern at an individual level. But we are now able to estimate what they add up to at a population level, and what they add up to is a lot: seven million early deaths every year, from a wide range of sources including cars and the humble domestic stove. In contrast, we've gone on to look at waste incineration, as one of the most notorious sources of air pollution, in a bit of detail. Here, the evidence shows us that these days, there are no longer any significant health effects for us to worry about. You probably want to know a bit more about where some of that evidence on air quality impacts of cars, stoves and power stations comes from, don't you? Thought so: in that case, read on.

CHAPTER 9

THE NUTS AND BOLTS

Now that you've read this far, it's time to take a deep breath and take a look at how we measure and predict air pollution in a bit more detail. Why should you care about how we get our data on air quality? I think it's important to know how robust, or otherwise, the data that we rely on for immensely important decisions are. Air quality is not by any means the only show in town, but it is one where important decisions need to be taken on the basis of technical data, so let's take the lid off the arcane world of air quality monitoring and modelling. And it's always fascinating when state-of-the-art science and technology meets the real world. We can produce air quality data to greater precision than almost any comparable discipline – more detailed than water quality or noise data, for example, and on a par with data relating to climate change. But all this precision can be a false comfort: our models are often accurate to within about 30% at best, and it gets a whole lot worse than that. Read on, if you dare, for horror stories of models under-predicting by a factor of ten.

ATMOSPHERIC CHEMISTRY, TAKE 1

One of the earliest theories of the atmosphere was proposed by Aristotle in the fourth century BC, and concerned the existence of something slightly weird called "the ether". The ether shouldn't be confused with the chemical unhelpfully known as "ether" or the group of chemicals known as "ethers" – just to be clear, ether exists, and ethers exist, but the ether doesn't, OK? The ether was thought to be the medium through which light travelled, but which was impervious to matter. This was a concept which proved surprisingly durable for over 2,000 years, despite having no grounding in reality at all. It wasn't finally disproved until James Michelson and Edward Morley's eponymous experiment (the "Michelson-Morley experiment") of 1887.

ATMOSPHERIC CHEMISTRY, TAKE 2

After the first faltering steps to introduce scientific methods, exemplified by Robert Boyle's book *The Sceptical Chymist* of 1661, the science of chemistry began to come of age in the eighteenth century. Experimental procedures improved, and scientists began to understand how atoms interact to form complex chemical compounds. The French nobleman and polymath Antoine Lavoisier was the first to bring some order to the chaos that had been left by hundreds of years of alchemy. He was ten years younger than a contemporary English minister and scientist, Joseph Priestley. Lavoisier's misfortune was to have been born into the French aristocracy at a time when that was about to become the most dangerous occupation going. Priestley's misfortune was to be wrong on one of the big scientific questions of the day.

Antoine Lavoisier took the opportunities afforded by an aristocratic background and education to develop his interests in biology, chemistry, geology, mathematics, meteorology and philosophy, so it was no surprise he went off to study law, and then became a tax collector. Law? Tax collecting? Come on, Antoine, you're better than that. Fortunately for the future of science, and atmospheric chemistry in particular, he devoted his spare time to some fairly breathtaking scientific breakthroughs. He developed a list of the elements as they were then understood, which included oxygen, nitrogen, hydrogen, phosphorus, mercury, zinc and sulphur. This was in the first chemistry textbook ever written (so anyone struggling to get to grips with GCSE chemistry – he's the man to blame), which is available to read today.[105] Most critically for the future of chemistry, Lavoisier recognised that oxygen and hydrogen were elements in their own right, and rejected the "phlogiston" theory, which had been around for about a hundred years.

PHLOGGING A DEAD HORSE

For much of the eighteenth century, the prevailing wisdom was that the process of combustion removed a substance called "phlogiston" from a combustible material. Thus, combustion was equivalent to "dephlogistication", and a substance which burned easily could be described as "phlogisticated" – full of phlogiston, which was ready and waiting to be removed by fire. Plants were thought to absorb phlogiston in a kind of reverse combustion, which explained why wood was so usefully combustible: it was all that phlogiston it had absorbed when it was a tree. Lavoisier realised that this theory was back to front. Combustion didn't remove something, combustion was a process of adding something, a something that he identified as oxygen. It's not that plants absorb phlogiston as they grow – no, they absorb carbon dioxide during photosynthesis, create energy-rich sugars, and release oxygen into the atmosphere as a waste product. The release of oxygen allows plants and trees to form sugar-based organic compounds containing carbon, hydrogen and oxygen. Plant structures (including wood) are mainly made up of cellulose, hemicellulose and lignin. Once dried to remove water, these organic chemicals burn very satisfactorily, absorbing oxygen and releasing a lot of energy in the process, to end up with carbon dioxide and water.

Unfortunately for Joseph Priestley's ongoing contribution to science, he was a devotee of the phlogiston theory. This conviction led to some strange conclusions – for example, rather than identifying oxygen as a substance, it was described as "dephlogisticated air" – that is, air that's just itching to take phlogiston out of a combustible material. Priestley was (probably) the first scientist to isolate oxygen in 1774, but unfortunately he didn't fully appreciate what he was dealing with. That gap was filled by Lavoisier and the Swedish scientist Carl Scheele. Scheele, a master of discovering new elements slightly too late, was also among the first to discover and identify nitrogen in the atmosphere in 1772, at about the same time as Henry Cavendish and Daniel Rutherford. As well as identifying oxygen and nitrogen, "hard luck" Scheele (as he was called by Isaac Asimov) independently discovered barium, molybdenum, tungsten and chlorine, and knew that manganese was out there – but didn't get the credit for his work.

THE FOUNDATIONS OF ATMOSPHERIC CHEMISTRY

Priestley continued to work on his theories and experiments in the new science of chemistry, as part of his broader interest in natural philosophy and as a founder of the Unitarian church. Meanwhile, Lavoisier developed the theory and practice of chemistry, carrying out some of the first quantitative experiments to show that the total quantity of matter is not changed during a chemical reaction – the principle of "conservation of mass". Oh and by the way, he took the first steps towards the periodic table of chemical elements, and invented chemical nomenclature and the metric system. Who knows what else he might have achieved, but unfortunately, in 1794 at the age of fifty, his head was cut off by the revolutionary government, nominally for tax reasons (Lavoisier had been a tax collector for the *Ancien Régime*). To be fair, he was pardoned a year and a half later, but that was about a year and a half too late to enable him to continue his work. Priestley also fell foul of the French Revolution – his support for the revolutions in France and America was so unpopular that his house was burned down and he eventually had to flee with his family to North America.

So the foundations of chemistry had been laid, and by 1774, while chemistry was still in its infancy, natural philosophers had already identified 99% of the atmosphere – nitrogen and oxygen. Lavoisier suggested the name "azote" for nitrogen, meaning "no life", because nitrogen is an asphyxiant gas. The name "azote" is used in many languages, notably French, and seems a better name than the more prosaic "nitrogen" which means "generating nitre" – nitre is another name for the mineral saltpetre, or potassium nitrate. Better than both is the German name for nitrogen, Stickstoff. No, it doesn't mean sticky stuff, which it manifestly isn't: the German name means "the stuff which chokes you" – the same concept as Lavoisier's word, azote. Oxygen was named "fire air" by Scheele, and originally referred to as "vital air" by Lavoisier. Both of these are more dynamic names than "oxygen" which means, rather inaccurately, "acid-forming". This was Lavoisier's idea, because he mistakenly thought for a while that oxygen was a constituent of every acid. By the time it became clear that oxygen isn't in every acid (with the benefit of hindsight, hydrochloric acid leaps to mind), it was too late to rename oxygen.

Even this basic understanding of what's in the air came too late for John Evelyn, our seventeenth century air quality manager from Chapter 6, as he was grappling with the problems of urban air pollution. All he could do was use common sense to look at where the pollution was coming from, and come up with a plan for shifting it somewhere else. Social equity was less of a problem for seventeenth century courtiers, so it was fine to suggest moving industrial activity to poorer areas where it wouldn't affect the nobility. Back then, and even more so during the industrial revolution, particulate matter was produced by coal burning in homes and industrial sources as well as agriculture, with the result that smoke, dust and airborne particles were an inescapable fact of life for everyone in a way that is hard to imagine nowadays. Historian Sean Adams quotes James Parton, a journalist who wrote in 1867: *"Smoke pervades every house in Cincinnati … It begrimes the carpets, blackens the curtains, soils the paint and worries the ladies."*[106] Attitudes to the grimy atmosphere of nineteenth century cities were ambivalent: while undoubtedly a nuisance, smoke represented evidence of honest endeavour and progress, and was thought to be beneficial for health. It isn't. But the good news for worried ladies is that nowadays, we don't burn very much coal in our towns and cities. Instead, the twenty-first century version of smoke, fine particulate matter ($PM_{2.5}$) comes from a very wide range of natural and man-made sources.

VARIATIONS FROM PLACE TO PLACE

The variety of sources of $PM_{2.5}$ means that levels are spread quite evenly across the whole of the UK. As we saw, the annual average levels of $PM_{2.5}$ measured in the UK in 2016 varied from 3 $\mu g/m^3$ in a very rural area to 17 $\mu g/m^3$ in central Birmingham – so a factor of about six between the highest and lowest measured levels. In contrast, oxides of nitrogen (NO_x) are produced mainly from combustion processes. There aren't many natural sources of NO_x, but there are plenty of sources arising from human activity – traffic, central heating and cooking in homes, boilers in commercial properties and public institutions (schools, hospitals, swimming pools and so on), using fuels in industrial processes, and generating electricity. These sources tend to be focused mainly in towns and cities, and also occur close to ground level, and so we'd expect to find a much greater variation in levels of NO_x. Sure enough, in 2016, the lowest level measured at any of the 147 UK national network monitoring stations was 2.7 $\mu g/m^3$ at Eskdalemuir in the Scottish borders. And the highest level was more than a hundred times higher, 297 $\mu g/m^3$ measured at Marylebone Road in central London (for air quality geeks, that's not an hourly or daily mean concentration – that's the annual mean concentration of NO_x at Marylebone Road). Nitrogen dioxide, one of the two components of NO_x, varied from 2.0 $\mu g/m^3$ at Eskdalemuir to 89 $\mu g/m^3$ in central London. The Marylebone Road monitoring station is stuck right next to the road, at a location where no one is continually exposed to the high levels of air pollution, so it doesn't necessarily matter too much how high the pollution levels are there. But it does illustrate how much levels of a man-made pollutant like oxides of nitrogen can vary from place to place.

Marylebone Road kerbside air monitoring station in central London[107]

Sulphur dioxide levels measured at a much smaller number of sites in 2016 varied between 0.6 $\mu g/m^3$ and 7.7 $\mu g/m^3$ – a factor of fourteen, much less variable than oxides of nitrogen. While sulphur dioxide is, like NO_x, emitted from man-made sources, these tend to be mainly industrial and commercial combustion processes, which discharge from chimneys at a high level above ground. By the time sulphur dioxide from these sources has reached ground level, the environmental concentrations are relatively low, and so there is less variation between measured concentrations. Also, what's left of the UK's sulphur dioxide monitoring network is mainly focused on urban centres, rather than on the areas close to coal and oil-fired power stations, which are likely to result in the highest levels of sulphur dioxide, so maybe they didn't record the highest levels which occurred in 2016.

Our air quality and environmental regulation policies are principally built around compliance with air quality standards and guidelines. That means that those responsible for managing air quality – national and local government officers, industrial process operators and regulators, and those responsible for new residential, commercial and infrastructure developments – need to know about the variation in levels of air pollution. For air quality management, we try to understand where the areas are, within which levels of air pollution could exceed the air quality standards. In the case of nitrogen dioxide for example, measured levels of nitrogen dioxide in 2016 were between 2.0 $\mu g/m^3$ and 89 $\mu g/m^3$, while the air quality standard is 40 $\mu g/m^3$. That's one heck of a variation, particularly when you remember that the UK can be taken to court and fined if we don't achieve the standard of 40 $\mu g/m^3$. At least, we can until we Brexit our way out of these pesky legal obligations.

Information from the UK's air quality monitoring network is mainly designed to inform an assessment of compliance with air quality standards and guidelines. But when we're looking at the impact of industrial processes or new developments on air quality, we have a slightly different objective. The aim then is usually to understand baseline air quality in the vicinity of the source of interest – that is, the levels of air pollutants that would occur in the absence of the industrial process or new development. We can then work out the additional contribution from the new or existing source, and assess that against the standard or guideline for an acceptable level of air pollutants. Evaluating baseline air quality is not necessarily straightforward, and a lot of air quality studies fall down on this fairly fundamental step. Part of the reason for this is the variability in baseline levels. So if you're looking at a development which could affect air quality over an area with a residential area, a town centre, an industrial estate and a motorway junction for example, you are likely to find that baseline air quality will be very different in these four different zones. Even within a zone such as a city centre, levels of air pollution might be fine in most places, but substantially higher in, say, a narrow, congested street with tall buildings. This variation needs to be reflected in an air quality assessment, which can add to the complexity of the work that's needed. No one said that an air quality specialist's life was easy.

AIR QUALITY MODELLING

So – how do we go about understanding levels of air pollution in such a complicated world? In two words, the answer is: dispersion modelling. Well, it's not the whole answer, but it's such a useful (and, in the wrong hands, dangerous) tool for managing air quality, that I'm prepared to risk an over-simplification.

A dispersion model is a means of estimating how stuff released from a source of pollution will disperse in the atmosphere. Any pollution emitted from any source into the atmosphere travels downwind, diluting in the atmosphere as it goes, so that the further away you get from the source, the lower the concentrations of stuff become.

We have a pretty good understanding of how this process works. The classic illustration of this process for pollutants emitted from a chimney stack looks like the figure on the next page.

Here, the average concentrations of pollutants downwind of the source are represented using the oval shapes. The wind is blowing along the axis labelled "x", taking pollution emitted from the source shown as a chimney in the positive direction along this axis. The average concentrations are highest at the very centre of the oval shapes, along the line labelled "Plume

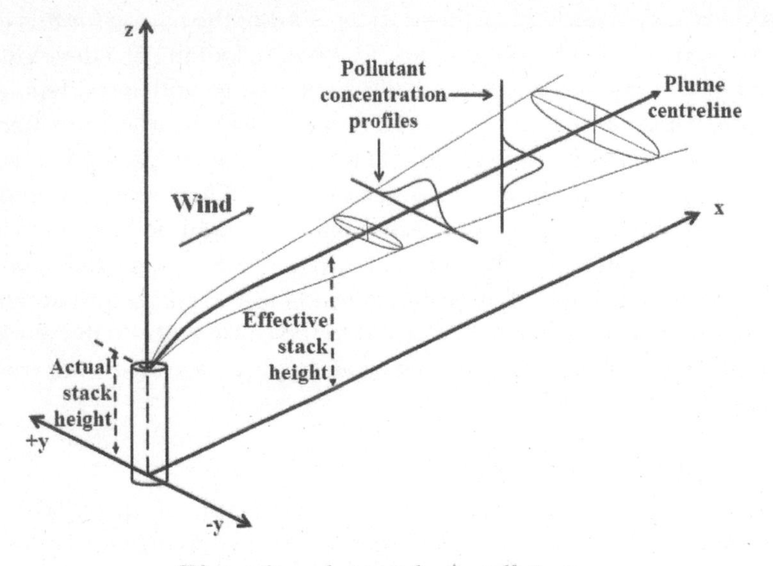

Dispersion of atmospheric pollutants from a chimney stack[108]

centreline". As the plume moves downwind, the concentrations get lower and lower.

Concentrations also get lower as you move away from this centreline both vertically (in the direction marked "z") and horizontally (in the direction marked "+y" and "-y") – that is, perpendicular to the wind direction. Concentrations are lower in both the positive and negative horizontal y-directions, and the same is also true in the vertical direction – concentrations are lower both above and below the plume centreline. This is illustrated by the little "pollutant concentration profiles" shown in the drawing.

THE GAUSSIAN MODEL FOR BEGINNERS

The shape of these little graphs is known as a Gaussian function, after the remarkable mathematician and (if he hadn't been born a century too late) Renaissance man Carl Friedrich Gauss. The little curvy graphs come from work carried out by Gauss on statistical distributions, which fortunately don't need to concern us for now. Suffice it to say that Gauss has a Wikipedia page just dedicated to the things that are named after him. This page lists over a hundred eponyms, including scientific discoveries, methods, laws and inventions as well as prizes, buildings, geographical features and a "student study space".

A short pause to recall that the Gaussian concentration profiles shown in the diagram are *average* values. At any one time, stuff released from the source is scattered across the downwind, cross-wind and vertical directions. When you average the concentrations, you end up with the little profiles, with highest concentrations near the centre of the downwind direction, and getting progressively lower the further downwind and away from the centreline you go.

A Gaussian curve is characterised by a mean and a standard deviation. The standard deviation of a Gaussian curve describes how pointy or flat the curve is. A Gaussian curve with a small value for the standard deviation looks like a sharp needle – the thin line in the graph. Conversely, a Gaussian curve with a large value for standard deviation looks like a cross section of a flat pancake – the thick line in the next graph – with concentrations decreasing only slowly away from the centre line.

THE IMPORTANCE OF STABILITY

What's the application to dispersion modelling? Well, when the atmosphere is very calm and still, a plume from an individual source does not spread very rapidly. The concentration along the plume centreline is pretty high (represented by the thin line in the graph with a standard deviation (σ) of 2), and as you move away from the centreline, the concentration quickly drops until it is indistinguishable from zero. Conversely, when the atmosphere is mixing vigorously, the pollutants spread away from the centreline very rapidly (represented by the thick line

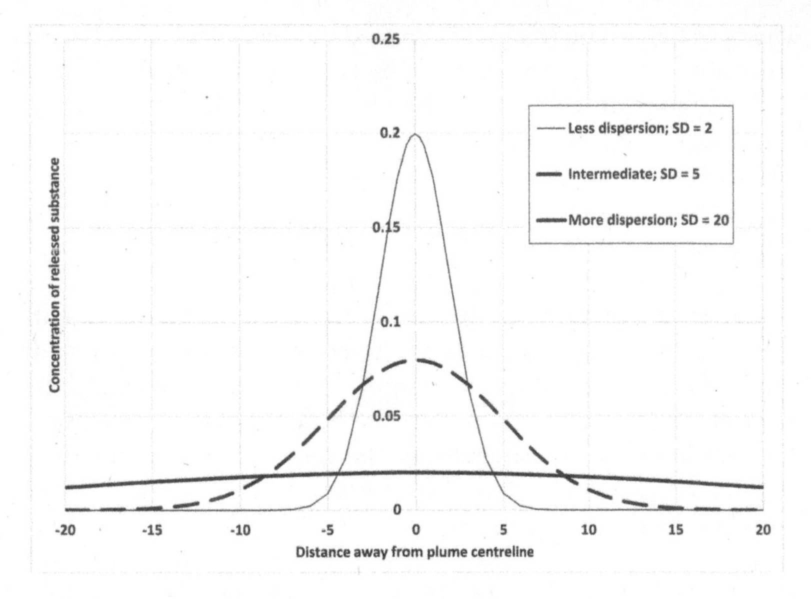

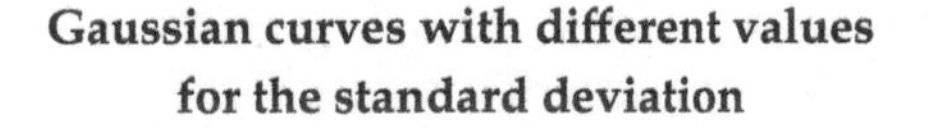
Gaussian curves with different values for the standard deviation

with standard deviation $\sigma = 20$). The highest concentration is still along the plume centreline, but it's much lower, and the pollutant levels are above zero a lot further away from the plume centreline.

How does this look in real life? The first photograph shows dispersion under calm, stable atmospheric conditions. Here, the plume does not disperse much, so that concentrations along the plume centreline will be relatively high, and concentrations away from the centre of the plume are very low – essentially zero, once you are any distance away from the plume (including at ground level close to the stack). The second photograph shows dispersion under unstable atmospheric conditions, with vigorous mixing of the plume in the vertical direction. We can't really tell how much the plume is mixing in the crosswind direction from this photograph (because it's taken in a crosswind direction from the source), but it would

Dispersion under stable and unstable atmospheric conditions[109]

certainly have a high σ_z value, representing rapid mixing in the vertical direction. Under these conditions, there would be significant concentrations at ground level even quite close to the source.

If we know the release rate of a substance, the wind speed, and the standard deviations which represent dispersion in the crosswind and vertical directions, we can work out the average concentration of a released substance at any point in space. This concentration is relevant to a particular set of weather conditions represented by the wind speed and the two standard deviation values (referred to as "sigma-y," σ_y, and "sigma-z," σ_z). It's not exact, because we never have exact data on weather conditions, and we also have to deduce the standard deviation values from meteorological observations and this is by no means an exact science. Also, this whole approach to modelling dispersion is only an approximation to what happens in reality. But having said that, it's not a bad approximation in many circumstances.

THE GAUSSIAN MODEL IN A BIT MORE DETAIL

Here's the basic equation which represents Gaussian dispersion, which is to be ignored by anyone who doesn't like maths, although if that's you, well done for getting this far and please feel free to hop forward a few paragraphs. I promise, it's the only equation in the book, but as equations go, it's a good 'un.

$$c = \frac{Q}{2\pi u \sigma_y \sigma_z} \exp\left(-\frac{1}{2}\frac{y^2}{\sigma_y^2}\right) \exp\left[-\frac{1}{2}\frac{(z-h)^2}{\sigma_z^2}\right]$$

Where "exp(*stuff*)" means "e to the power of *stuff*", and e is a numerical constant with the value 2.718281828.

c is the concentration of a released substance at a point x metres downwind of the source, y metres away from the wind direction in a perpendicular direction, and z metres above ground level *(x, y, z)*

Q is the release rate of the substance – say in grams per second

σ_y and σ_z are the standard deviations describing the extent of dispersion in the crosswind and vertical directions respectively

h is the height of the plume centreline

Without going into this rather beautiful equation in detail, there are a few hopefully interesting points which drop out of the equation. Firstly, we can simplify matters by looking at concentrations along the plume centreline; that's the line where $y=0$ and $z=h$. We can put these values into the equation, and find that along the centreline, both the terms exp(*stuff*) become 1, and so the equation simplifies to the release rate divided by 2π, the wind speed, σ_y and σ_z. As soon as you move off the plume centreline, y is no longer 0 and/or z is no longer equal to the plume height *h*. Because y^2 and $(z-h)^2$ are now greater than zero, the terms exp(*stuff*) become less than 1 – that is, the concentration gets lower as you move away from the plume centreline. That makes sense – the highest concentrations should indeed be directly downwind from the source, on the plume centreline.

Secondly, other factors being equal, the concentration of a given substance is directly proportional to the release rate and inversely proportional to the wind speed. The more stuff that is emitted, the higher the resulting concentration, which seems reasonable. And the faster the wind takes the released substances away, the lower the concentration – again, pretty much what you'd expect. However, this does bring us on to one of the difficulties with using the Gaussian dispersion equation. When the wind speed is zero, the equation falls down – it predicts that concentrations are infinite. Well, this isn't really the case: if the wind speed really is zero, then emissions from a source would go straight up, or just stay where they are. In practice, wind speeds are rarely if ever absolutely zero. However, using the Gaussian equation to represent dispersion does become less reliable at low wind speeds. This is partly the nature of the atmospheric dispersion process: under low wind speeds, concentrations can vary dramatically depending on the wind speed and direction. A small difference in wind speed can result in a substantial change in concentrations. A further complicating factor is that the weather observations that we use when applying this approach are usually taken from airports. Airport meteorologists are often very interested in high wind speeds, for good reasons of aircraft safety, but not very interested in low wind speeds. Conversely, air pollution levels tend to be at their highest when wind speeds are low. I've never heard anyone bellowing during a gale, "I'm really concerned about the excessive levels of airborne nitrogen dioxide," whereas I have once landed sideways at Aberdeen airport because of the gale force crosswinds, which underlined the importance of high wind speeds to me very firmly. *Welcome to Scotland*. Because airport operators are primarily concerned about moderate and strong winds, the equipment used to measure wind speeds at airports is often designed to give the most reliable data for high wind speeds, and may not be very good for the lower wind speeds. All we can do is discount weather observations which have unreliable data for low wind speeds – below say half a metre per second (that's one mile per hour, or maybe up to 1 on the Beaufort scale, "Wind direction shown by smoke drift but not by wind vanes" – which kind of illustrates my point about the possible unreliability of weather measurements under low wind speed conditions). We then need to acknowledge this shortcoming of Gaussian dispersion models when interpreting model results.

This is a particularly big deal for sources at ground level, of which the classic example is, of course, road traffic. Because individuals might be located close to the exhaust pipes in question, the local traffic emissions are a big component of their exposure to air pollution. As a result, uncertainties in how these emissions disperse might make a big difference to the levels of pollution the individuals in question are exposed to. On the other hand, the whole point of an industrial chimney is to take the pollution away from the people. So even if you're equally uncertain about how the pollution from a chimney stack might disperse, the effect on any individual down at ground level is typically a lot less.

And thirdly – the values for the standard deviation in the crosswind and vertical directions (σ_y and σ_z) are hardwired into this equation. You've got to know what they are, otherwise you can't use the equation. The sigma values depend on the distance downwind from the source: the further downwind you are, the greater the degree of spread, and hence, the greater the sigma-y and sigma-z values. It's important to get these values as accurate as possible, because any systematic errors are likely to result in significant deviations in the model results. Unfortunately, you can't go out and buy a "sigma-ometer" to measure the standard deviation of dispersion. What you have to do is develop an algorithm – that's a list of instructions like a cooking recipe – which uses widely available meteorological observations to calculate values for σ_y and σ_z. So that's what the people who develop air quality models have done. The algorithms which are used in current dispersion models are derived from some fairly venerable experiments – field trials, where tracer substances were released into the atmosphere and downwind concentrations measured along with the weather conditions, and wind tunnel experiments where the meteorological conditions could be controlled more reliably.

USING THE GAUSSIAN MODEL

So the one and only equation in this book (not counting chemical formulae, which are hardly equations) is the engine for computer-based dispersion modelling. All a computer modelling study does is to apply some refinements to this basic picture of dispersion, and run it lots and lots of times to reflect all the different weather conditions that are likely to affect the source. The model brings together information on the source, a set of weather conditions and a location where concentrations are to be calculated (often called a "receptor", a term I don't like because of its implication that the role of people living near to a source is just to receive the pollution doled out to them – however, I can't deny that it's a useful shorthand). The model works out the dispersion parameters which apply to emissions from the source at the receptor under the set of weather conditions, and then uses these to calculate the concentration of pollutants emitted from the source at the receptor location. The calculation is then repeated to evaluate the concentration of every substance emitted from every source in the study at every receptor. It's tedious, but of course, tedious, repetitive calculations are exactly what computers are good at.

We don't know what the weather conditions will be at any specific time in the future, but we do often have pretty good information on how the weather conditions have been in the past. So what we do is to run the model using a set of observations from some reasonably representative weather station, over a reasonably representative period – say, observations made every hour for a year, or five years. Running the model using a five year meteorological dataset typically gives us over 40,000 modelled concentrations of every pollutant emitted from every source, at every location that we're interested in. The model stores these modelled concentrations and processes them to give descriptive values – for example, the average (mean) concentration, or the median value, or the maximum calculated value for any of the 40,000-plus sets of weather conditions.

This averaging process is a real strength of the computer modelling approach. Individual modelled concentrations for specific sets of weather conditions are subject to all kinds of uncertainty – how representative was the weather data? How well does the model calculate the dispersion standard deviation values based on the meteorological observations? Do we know that the pollutant release rate was appropriate for that specific period? This makes trying to model concentrations under particular conditions fraught with difficulty. Often, air quality model results for individual sets of circumstances (for example, trying to work out which sources might have contributed to a strong smell which occurred at a specific time) aren't worth the paper they are written on. In contrast, when we have say 40,000 calculated concentrations for different sets of weather conditions, uncertainties in individual values are much less important. Indeed, the averaging process is so good that, under ideal conditions, we can use a computer model to calculate annual mean concentrations due to emissions from a single, isolated point source to an accuracy of – wait for it, this is going to be good – about 30%. 30%? That might not sound very accurate to anyone who's used to dealing with better controlled systems – laboratory experiments, or engineering design for example. But in the complex, messy real world where environmental scientists try to ply their trade, that's good going. Because we can only hope to be accurate to about plus or minus 30% on a good day, it does make me cross when my fellow dispersion modellers either deliberately or accidentally try to give their results some kind of spurious scientific validity by quoting results to lots and lots of significant figures. For example, I've just seen a report which gives a modelled concentration of nitrogen dioxide of 634.5 micrograms per cubic metre. As if we can be confident that it isn't 634.4 or 634.6. We can't be any more accurate than "about 600", or at a push "about 630". Modelling is just not that accurate.

COMPLICATIONS

I did say that we can get within about 30% under ideal conditions. Pretty soon, however, we start encountering sources which don't conform to the ideal situation shown in the diagram. Not every source of air pollution is an isolated chimney in a flat field.

Sticking with chimneys for now, one of the biggest issues that affects dispersion is the presence of buildings close to the stack. The whole point of a chimney is to take air pollutants away from the ground, to allow them to disperse to lower concentrations. As you will know if you've ever walked past a large building in a strong wind, or even quite a gentle breeze, buildings interrupt the wind flow. This results in a turbulent zone in the air around the building. If a chimney is built close to a building and too low compared to the height of the building, emissions from the chimney can easily be caught up in the turbulent zone and dragged down to ground level before they have a chance to disperse properly. That can defeat the object of having the chimney in the first place. Even worse, sometimes chimneys are located adjacent to buildings such as residential tower blocks or hospital wards. Putting a source of pollution at an insufficient height adjacent to such a building is asking for trouble: the building occupants run the risk of experiencing more or less undiluted emissions from the chimney. There is some pretty simple guidance explaining how to avoid this unfortunate situation, but it's not always properly followed.

Fortunately, we have a good understanding of the atmospheric processes which go on when winds interact with buildings, in particular from experiments in wind tunnels. This enables us to take account of these effects when assessing emissions from sources of pollution. It is a bit of a trade-off, however: the models are good at allowing us to take account of building effects, but this does introduce some additional uncertainty into the modelling process. Better to avoid building effects where possible, but of course it's not always possible, and often we're dealing with existing sources of emissions where the chimney and building are already in place.

We also have to take account of the characteristics of the release. We've thought about the release height, but in fact, we're more interested in the *effective* release height. All releases from chimneys come with a bit of momentum – that's the zip with which they are pushed out of the chimney. The momentum is related to the rate at which the total quantity of gases are released from the chimney, and the speed of the release. A release with reasonable momentum will rise a bit in the atmosphere on release, until its momentum is dissipated. You can actually get a small dispersion bonus by reducing the diameter of the stack so that the same amount of gases are released at a faster velocity. This can give a small benefit in reducing the highest ground-level concentrations, although it doesn't quite come free. The process needs to work a bit harder to push the gases out through a more restrictive flue, so there's likely to be an energy cost and possibly a need for a bigger fan to give the gases the required oomph to get up the chimney.

The other factor which affects the effective release height is the temperature. Emissions from many industrial processes which involve combustion are released at high temperature. An emission above the ambient temperature also results in a rise in the plume on release. This increase in the effective release height also gives a benefit in terms of dispersion and reduced ground-level concentrations. These days, we often try to make processes more efficient by using the heat in the flue gases from combustion processes before releasing them into the atmosphere – most commonly, in plant which generates both useful heat and useful electricity, referred to as "Combined Heat and Power" or CHP. That makes all kinds of sense, of course, but it does mean that the gases come out of the chimney both cooler and slower, both of which are bad news for the effect of the process on local air quality. It's pretty straightforward to make sure this doesn't cause a problem in practice, but when added up, a lot of CHP in a large urban area might actually be quite bad news for air quality in the city.

Once we get away from chimneys and start trying to model concentrations due to emissions from other kinds of source, we have to deal with a load more uncertainties. Road traffic is the number one source of plenty of pollutants in many areas. As well as traffic which is already on the roads and affecting air pollution experienced by local people, many new developments, such as residential or commercial properties, affect traffic flows in the local area. Why are vehicle emissions so important? It's a combination of the number of vehicles on the road, the rate at which each vehicle emits pollutants, and the proximity of the source to receptors. I can't help noticing that vehicle exhausts are conveniently placed at about the same height as a toddler sitting in a pushchair. I'm sure this juxtaposition isn't deliberate, but it does mean that we are often very interested in modelling the levels of air pollution from road traffic. And releases from vehicle exhausts have very different characteristics to releases from chimneys.

MODELLING TRAFFIC EMISSIONS

In principle, we use exactly the same approach to modelling emissions from vehicles as we do from chimneys. The same principles apply, with the released substances transported downwind, while the plume spreads in the crosswind and vertical directions. We don't have the same standardised datasets from field experiments to enable us to verify the model performance, so when modelling traffic emissions in the UK, we try to apply a local correction to the models. We do this by carrying out the modelling study to estimate concentrations of the pollutants of interest (usually nitrogen dioxide) at one or more locations where monitoring has also been carried out. By comparing the measured levels to the modelled levels, we can work out a correction factor which can be applied to the model results to reproduce the measured concentrations.

Interestingly, when we carry out this process, we find that the uncorrected air quality models often underestimate the measured levels of nitrogen dioxide. Correction factors of two or three are not unusual. This observation was one of the pointers which suggested that databases of vehicle emissions of oxides of nitrogen were under-representing emissions, in the way that you might expect if, say, there was widespread use of defeat devices by a vehicle manufacturer. Other contributory factors to model under-prediction include the difficulty of using models under low wind speeds. As we saw, this is more of a problem for ground level sources like road traffic than it is for modelling emissions from factory chimneys, so it is often relevant for studies of urban air quality where the focus is mainly on emissions from low level sources like road traffic.

While correction factors of two or three are common, I have seen studies which required the model results to be "adjusted" by a factor of nine or ten: putting it another way, in these cases, the model was more or less useless. The larger correction factors were usually calculated in cases where the model was being used to calculate pollution levels in a highly congested city centre situation, with many tall buildings interrupting air flows in unpredictable ways. The modelling studies may also have underestimated vehicle emissions when driven under congested stop-start conditions. Under these circumstances, we shouldn't just multiply the model results by ten – we should go back to the drawing board to look at what was going wrong with the model and either fix it, or give up the attempt to use a dispersion model, and try a different approach.

But by and large, using this model adjustment approach generally allows us to use dispersion models to reproduce current levels of air pollutants due to traffic emissions. Then we can start to get the real benefit of a dispersion model, which is to look at the air quality impacts of future developments and interventions. Will a new development cause a new air quality problem? If we change traffic arrangements around a polluted city centre, will that improve air quality? If we replace a city's bus fleet with low- or zero-emitting vehicles, what benefit will that have on levels of nitrogen dioxide and $PM_{2.5}$? What effect would a Low Emission Zone have on city centre air quality – what vehicles should be covered, and how much should drivers be charged in order to ensure that air quality standards will be achieved? These are important questions which need answers to ensure that everyone has access to clean air. Atmospheric dispersion modelling has (some of) the answers, which perhaps begs the question why we're still some way off achieving the air quality standards in numerous urban areas in the UK, some twenty years after the process of local air quality management got started in the UK and Europe. There are a lot of reasons for this, mainly to do with the costs and unpopularity of measures that would be needed to finally nail the standards throughout the country. After two decades of air quality modelling and assessment studies, I don't think we can really say that we don't know how to achieve the air quality standards. We know what we have to do, we are making progress, and we know what's left to do: but we're not there yet.

THE GAUSS LEGACY AND BEYOND

So this is my old friend the Gaussian dispersion model, which has been a workhorse for air quality scientists for many years. The origin of using Gaussian models to represent atmospheric dispersion is a 1936 publication by Bosanquet and Pearson entitled *The Spread of Smoke and Gases from Chimneys.*[110] This theoretical paper was followed up in 1947 with an analysis based on experimental data from the Porton Down research station, published by Graham Sutton.[111] These highly technical papers entered the realms of practical application with the characterisation of atmospheric stability developed by Frank Pasquill in 1961. Pasquill's innovation was to define six (later, seven) classes of atmospheric stability. These ranged from very unstable atmospheres through neutral conditions, to very stable conditions. Based on a relatively small number of measurements, it was possible to identify the atmospheric stability class. This enabled models to be developed in which σ_y and σ_z values were tabulated for each stability category. Together with the increasing access to computing power, air quality scientists were finally able to run dispersion models more or less routinely. Nowadays, most modelling studies no longer use the Pasquill stability categories. While they are a useful and practical way of describing atmospheric stability, more recent developments in atmospheric physics enable us to describe the structure of the atmosphere more reliably. And the original concept of dispersion from a factory chimney has been developed into specialist systems for modelling emissions from all kinds of sources – not just chimneys but also exhaust pipes, and sources scattered or dispersed over a wide area such as domestic coal fires, landfill sites or water surfaces. But the principles are the same: it all comes back to Johann Carl Friedrich Gauss, and his handy little formula.

Although I've been using Gaussian dispersion models for twenty-five years, and they are the basis of most air quality modelling studies, I have finally had to admit that other air pollution models are available. For example, the Community Multi-scale Air Quality (CMAQ) modelling system is widely used for regional air quality assessments where the source-specific nature of the Gaussian dispersion model is less useful. It is used as part of a wider package of models to represent the transport of materials through the atmosphere based on a set of boundary conditions, weather conditions and emissions. This kind of model, focusing on locations in space through which the atmosphere passes, is named Eulerian after another prolific eighteenth century mathematician, Leonhard Euler.

AIR QUALITY MEASUREMENTS

The best models in the world are no use without solid data on the actual levels of air pollution in the atmosphere to back them up. But measuring air quality is not that easy. As we saw in Chapter 3, air pollutants are present at microscopic levels in the atmosphere – typically around 0.000001% of the atmosphere.

DIFFUSION TUBES

In that context, one of the most amazing monitoring techniques out there is a simple little plastic tube, costing about ten pounds to produce and analyse. The inside of the tube is coated with a chemical called triethanolamine. All you have to do is fix the plastic tube to the side of a building or a lamp post for a month or so, and then send the tube off to a laboratory (with a tenner, of course).

A diffusion tube attached to a lamp post[112]

The laboratory can then analyse the residue in the tube, and tell you the average nitrogen dioxide concentration over the month that the tube was out on site. Considering how cheap and simple the technique is, the measurement is reasonably (not to say, surprisingly) accurate, especially when two or three tubes are used in the same place. These diffusion tubes are used by many local authorities to give a cheap and valuable indication of nitrogen dioxide locations across their boroughs. My own local authority carries out a programme of diffusion tube monitoring, with six of its diffusion tubes within a few hundred metres of my home recording nitrogen dioxide levels above the national air quality standard in 2016. Maybe I wasn't so clever buying a house on the east side of town after all... Combining diffusion tubes with targeted monitoring using more accurate techniques where needed, and modelling to extend the findings of monitoring surveys, gives a lot of valuable evidence to characterise air quality.

Diffusion tubes are both cheap and simple to use, which opens up the opportunity for wider involvement of local communities in air quality monitoring. Provided the tubes are located, handled and analysed correctly, community surveys can provide useful information on air quality in areas of concern to local communities. However, diffusion tubes aside, air quality monitoring tends to be quite an expensive and technical business. This tends to place a practical limit on community-commissioned and conducted air monitoring surveys. A community will often need to draw on the expertise of its local authority in order to ensure that an air quality survey is well designed and carried out.

MEASURING NITROGEN DIOXIDE

When I started working in air quality in 1992, there were just eighteen stations measuring nitrogen dioxide in the UK national network. Progress from the thirteen stations in 1990, but not unreasonably, the government was subject to some criticism that this level of monitoring wasn't enough. Pretty soon, air monitoring was made a requirement of local air quality management, and around the same time, the European Commission set minimum monitoring requirements in its air quality directives. That led to a dramatic expansion in the extent of air quality monitoring throughout the UK during the 1990s. Air monitoring continued to expand during the 2000s, but in recent years the monitoring programme has started to slow down, or maybe we could say it has a sharper focus, along with wide-ranging cuts in government spending since the late 2000s. Even so, there remain 136 monitoring stations in the national "automatic urban and rural network", enough to give us a good picture of air quality throughout the country. Even if there isn't a monitoring station near you, there is enough information to enable your local authority to determine whether air pollution is likely to be a problem or not.

Nitrogen dioxide levels are measured using an instrument which for many years has been known as a "NO_x box". The first step in the process is to generate ozone. The ozone is exposed to the air sample, and it reacts with nitric oxide (NO) in the air sample to form nitrogen dioxide (NO_2). But this is no ordinary nitrogen dioxide: this is nitrogen dioxide with a bit of extra energy. The nitrogen dioxide loses the energy as a little bit of light at a specific wavelength. This is chemical luminescence, or chemiluminescence for (slightly) short: never use two technical terms when you can combine them into one super-technical word ("Wortverknüpfung," as they say, appropriately enough, in German). By measuring the intensity of the light emitted by this energetic nitrogen dioxide, the concentration of nitric oxide in the air sample can be calculated. That's nitric oxide, so we haven't yet measured nitrogen dioxide, but we're getting there. The next step in the process is to pass the next air sample through a catalyst which converts all the nitrogen dioxide present to nitric oxide. This converted air sample is then passed through the same analytical process. Because the nitric oxide in the converted sample has come from both nitric oxide and nitrogen dioxide in the original air sample, the result of this second measurement is the total concentration of nitric oxide and nitrogen dioxide – the family of substances known as NO_x. So the instrument gives measurements alternately of nitric oxide and oxides of nitrogen. Averaging each of these over an hour gives the hourly mean NO and NO_x concentrations, and we can at last get to the nitrogen dioxide concentration by subtracting NO from NO_x.

As with many other monitoring instruments, the NO_x analyser is normally used to give hourly mean concentrations, although both longer and shorter averaging times can be used if appropriate. But air quality standards for nitrogen dioxide are set on the basis of hourly mean concentrations, and there hasn't in the past been much benefit in considering concentrations over shorter averaging periods – although that may be changing now that we have new and more efficient methods of storing, analysing and getting extra value out of our monitoring records. The instrument needs to be calibrated periodically so that any drift in measured concentrations can be determined and the concentrations adjusted accordingly. For measurements made as part of the UK's national network, there is a process of data verification designed to ensure that measurements are as complete, accurate and robust as possible.

INVESTIGATING NITROGEN DIOXIDE MEASUREMENTS

Here's the kind of information that a NO_x analyser gives us:

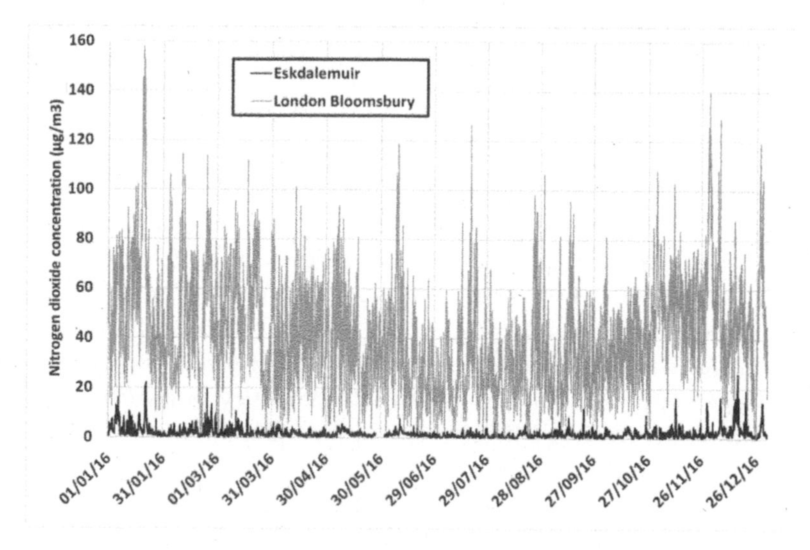

Nitrogen dioxide levels measured at a rural site and an urban site in 2016[113]

The graph shows how levels of nitrogen dioxide measured at an urban station and a rural station went up and down throughout

2016. There's clearly a lot of variation in nitrogen dioxide levels – something big is going on to make the levels vary so widely. During 2016, the hourly mean level measured at the London Bloomsbury site in Russell Square was 8 $\mu g/m^3$ at its lowest, and 322 $\mu g/m^3$, forty times higher, at its highest. And on 6 December, for example, the hourly mean level plummeted from 126 $\mu g/m^3$ during the hour between eleven and twelve in the morning, to 61 $\mu g/m^3$ the following hour, which was, of course, lunchtime. That's a massive change from one hour to the next. So what's going on to bring about such dramatic and rapid changes in nitrogen dioxide levels?

The advantage of having such detailed measurements is that we can look at the measured concentrations alongside other information (like weather conditions and traffic flows) to answer these kind of specific questions, and also analyse data very rapidly to confirm compliance (or otherwise) with air quality standards.

Let's take a quick look at the measured concentrations shown in the graph. The first question we're interested in is – did the measured concentrations comply with the air quality standards? That's pretty straightforward to start with: the air quality standard for annual average levels of nitrogen dioxide is 40 $\mu g/m^3$. The average concentration measured at Eskdalemuir (a small village in the Scottish borders) in 2016 was 2.0 $\mu g/m^3$, so complying with the air quality standard by a considerable margin. However, the average concentration measured at Bloomsbury (a swanky area of central London) in 2016 was 40.9 $\mu g/m^3$, just out of compliance with the air quality standard.

Then things get a little more complicated. There is also an air quality standard, which sets a limit of 200 $\mu g/m^3$ on hourly average nitrogen dioxide concentrations (the 40 $\mu g/m^3$ limit is on *annual* average concentrations). Pretty straightforward, except that nitrogen dioxide levels are allowed to exceed the limit value for eighteen hours per year. We usually interpret this by saying that eighteen hours per year represents 0.205% of the number of hours in a year. So nitrogen dioxide levels should be below 200 $\mu g/m^3$ for 100% of the year, minus the 0.2% of the year when levels can be as high as you like, and it won't affect whether or not the standard is exceeded. That is, nitrogen dioxide should be below 200 $\mu g/m^3$ for 99.8% of the year. The way that we check whether we've complied with the air quality standard is to analyse the measured concentrations to find out what the 99.8% highest concentration was. Going down this route enables us to decide whether we complied with the standard or not, and also to get an understanding of how close we were to exceeding the standard.

The 99.8th percentile of hourly mean nitrogen dioxide levels at Eskdalemuir in 2016 was 17 $\mu g/m^3$ – again, well within the limit value of 200 $\mu g/m^3$. Not much for the 265 residents of Eskdalemuir[114] to worry about there. And there is also good news for the residents of Bloomsbury – the 99.8th percentile concentration here was 133 $\mu g/m^3$, also complying with the air quality standard by quite a considerable margin. At 133 $\mu g/m^3$ of nitrogen dioxide or below, no one would be expected to suffer any acute ill-effects due to nitrogen dioxide. The standard is set at 200 $\mu g/m^3$ because the lowest level at which reliable laboratory experiments have identified a detectable effect on respiratory health in people already suffering from asthma is 375 $\mu g/m^3$. Applying a 50% margin of safety brings us to about 200 $\mu g/m^3$.[115] So there would not have been any ill-effects for the 99.8% of the time that levels were 133 $\mu g/m^3$ or lower – and indeed, it's unlikely that there would have been any noticeable effects for the remaining eighteen hours of 2016, when levels went up as high as 322 $\mu g/m^3$.

Let's have a quick look at one week during 2016.

Here, we can see nitrogen dioxide levels going up and down in response to daily and weekly events. Nitrogen dioxide levels are generally lower overnight at the Bloomsbury site compared to the levels during the day, for two reasons. Firstly, there is a lot less traffic around to emit oxides of nitrogen, and

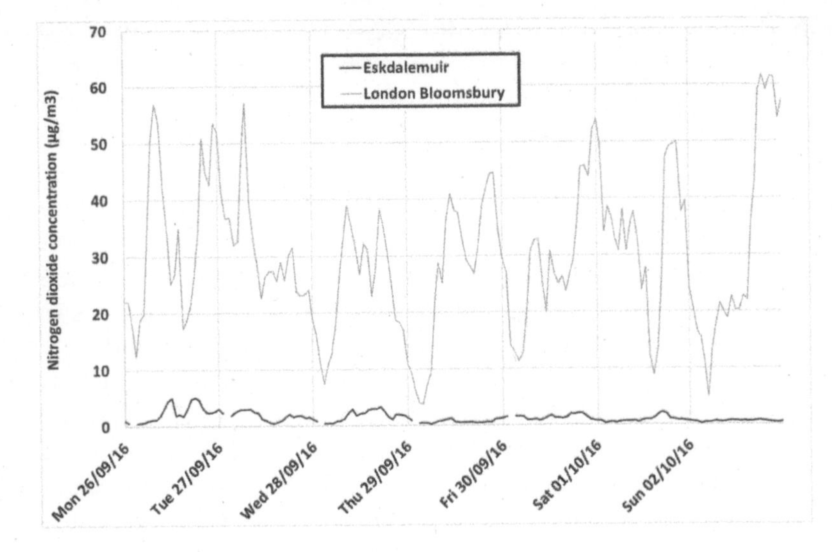

Nitrogen dioxide levels measured at a rural site and an urban site, 26 September to 2 October 2016[117]

secondly, there is no sunlight-driven oxidation of nitric oxide to nitrogen dioxide. On some days, we can see an increase in nitrogen dioxide levels in the morning – particularly Monday and Wednesday. You would expect to see a clearer indication of local traffic emissions if we were to look at total oxides of nitrogen rather than nitrogen dioxide, and if we look at a nearby monitoring station more directly affected by road traffic. So let's do just that – here are the measured levels of total oxides of nitrogen for the same period at the Marylebone Road kerbside monitoring station, which is on a much busier road just round the corner from Russell Square.

On the following page, we can see high oxides of nitrogen levels at around the busiest period of the morning rush hour on Monday, Tuesday, Wednesday and Friday. There is also a peak in levels of nitrogen dioxide during the evening rush hour on

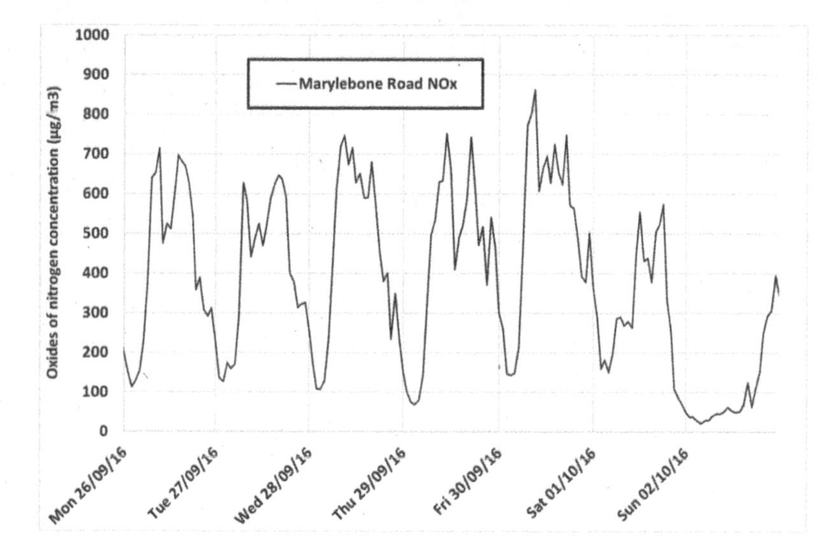

Oxides of nitrogen levels measured at Marylebone Road, 26 September to 2 October 2016[117]

Monday, Tuesday and Thursday. Levels of the traffic-indicator pollutant oxides of nitrogen were lower at the weekend than during the week, with the lowest recorded levels of NO_x during this week occurring on the Sunday morning.

That's just from a single week: we can get a better picture of hour by hour variations by looking at the levels averaged by hour of the day over every week of the year – here, showing data measured during weekdays (Monday to Friday) at Marylebone Road:

So on average, the highest levels occur during the period seven a.m. to ten a.m., when the highest traffic flows probably coincide with relatively low winds. Anyone who's been anywhere near Marylebone Road won't be surprised to see that air pollution levels here don't really drop very much during the middle of the working day – and that's because traffic continues more or less unabated throughout the day. Later on in the afternoon, we do

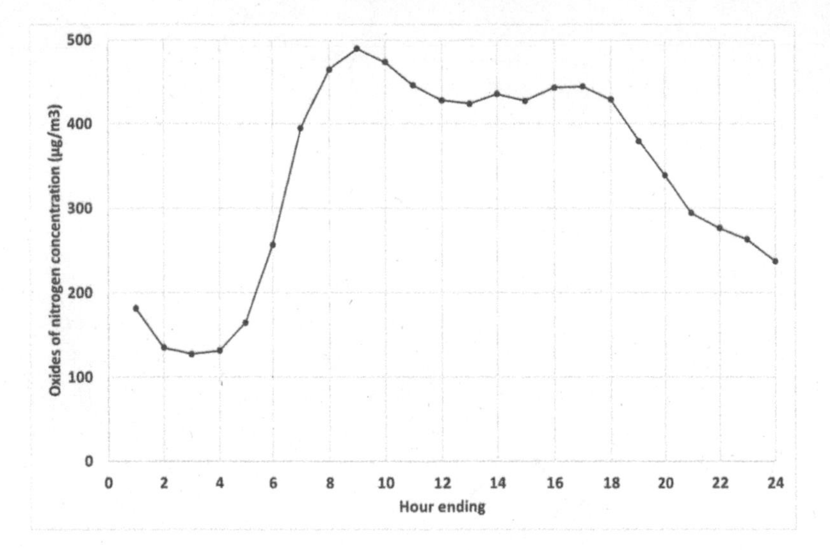

Oxides of nitrogen levels measured at Marylebone Road by hour of the day on weekdays during 2016[117]

see a smaller increase in measured levels of oxides of nitrogen, corresponding to the evening rush hour, before concentrations drop off during the evening and early morning.

Going back to the measured levels of nitrogen dioxide at Bloomsbury, the highest concentration of nitrogen dioxide, almost 160 $\mu g / m^3$ was measured during the second half of January. To be precise, the highest concentration was, perhaps surprisingly, recorded between two and three o'clock in the morning, on 20 January 2016. That's nothing to do with rush hour traffic, so let's have a look at this period in a bit more detail to understand what's going on:

Looking at a few days either side of this peak shows that there was a period of three days (19, 20 and 21 January) when levels of nitrogen dioxide were particularly high. An extended period of elevated pollution levels like this is not so much linked to local

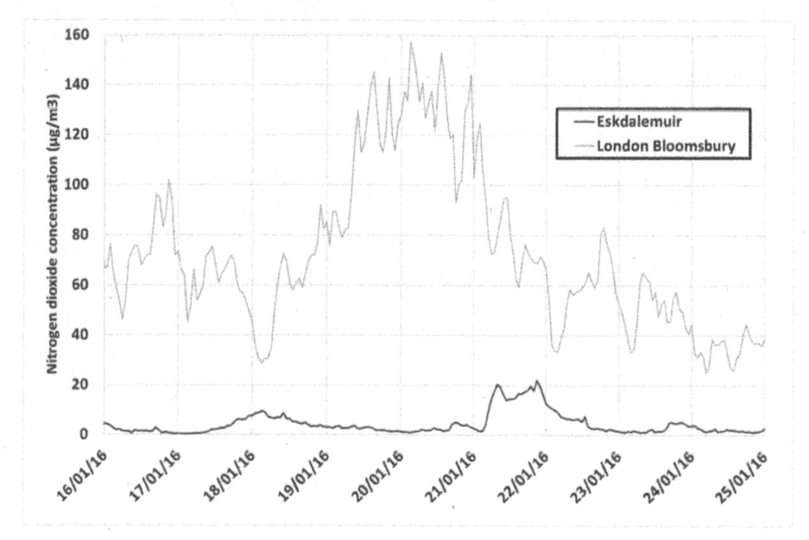

Nitrogen dioxide levels measured at London Bloomsbury and Eskdalemuir during a pollution episode[117]

sources of pollution but to the effect of unfavourable weather conditions on regional sources of pollution. During Tuesday 19 January and Wednesday 20 January, the wind speed in London was very low, around two to three kilometres per hour. Temperatures were also low, within a few degrees either side of freezing point before increasing above freezing during the morning of 21 January. Because of these unfavourable weather conditions, air pollution levels were elevated throughout London, returning to more normal levels once wind speeds and temperatures started to pick up. Even though the measured levels of nitrogen dioxide at Bloomsbury were on the high side, they did not go above the Defra descriptive air pollution index level for "low pollution". At the nearby Marylebone Road site, the measured levels did just scrape into the lowest "moderate pollution" band for a couple of hours during this episode.

Looking at the measured levels of nitrogen dioxide at Eskdalemuir during January 2016, there was a mini-pollution episode here as well, with nitrogen dioxide levels heading up to a not-very-scary 20.4 $\mu g/m^3$ on 21 January. This mini-peak in nitrogen dioxide levels is the tail end of the larger scale episode which affected much of the UK over these few days. Still, 20.4 $\mu g/m^3$ was the fourth highest nitrogen dioxide concentration recorded at Eskdalemuir in 2016, so it should still count as an air pollution episode relative to the levels normally experienced in that part of Scotland, even if it wasn't one that would have had any consequences for the health and well-being of the local community.

PICTURING NITROGEN DIOXIDE MEASUREMENTS

As well as looking at the numbers, we can analyse air quality and meteorological datasets using visual techniques – for example, the freely available "OpenAir" package. This system enables the factors which affect measured levels of air pollutants to be investigated, providing a valuable additional tool for developing measures to improve pollution levels. For example, we can take the levels of oxides of nitrogen measured at Bloomsbury in 2016, and investigate how they depended on wind speed and direction using something called a "bivariate plot".

The different shades in the plot represent the measured NO_x concentration for different combinations of wind speed and direction. Concentrations near the centre of the plot represent measurements taken under low wind speeds, and further out from the centre represents measurements at more rapid wind speeds, up to a decidedly blusterous twenty metres per second (seventy kilometres per hour) – that's a Force 8 Gale, for anyone who prefers the glorious imprecision of the Beaufort Scale. So the measured concentrations during the January episode, when wind speeds were only a few kilometres per hour, contribute to the very highest pollution levels represented by the dark colours

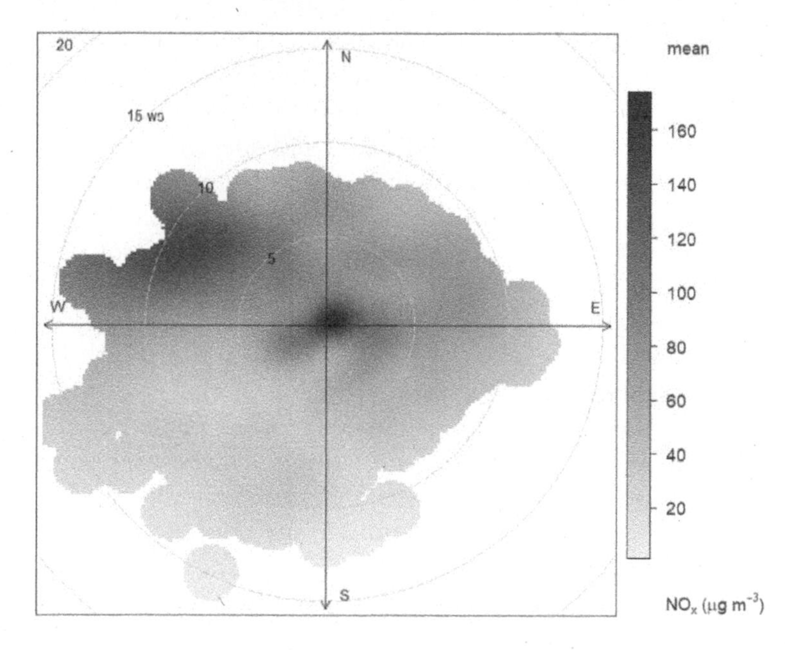

Bivariate plot for measured levels of NO_x at London Bloomsbury in 2016[116]

right in the centre of the plot. The location around the plot represents where the wind was blowing from. So, for example, concentrations in the top left hand quarter of the graph represent measurements made when the wind was blowing from the north-west.

What does this plot show us? Firstly, the high levels recorded under very low wind speed conditions show the contribution of local sources to measured levels of air pollutants at the Russell Square air quality monitor. As well as traffic on nearby roads, oxides of nitrogen emitted from other parts of the local area are likely to make a significant contribution to measured concentrations when wind speeds are very low – as we saw during the January 2016 episode. Secondly, it looks like there is a significant contribution to levels of oxides of nitrogen from a source located to the north-west of the monitoring station. This makes the greatest contribution to measured levels of oxides of nitrogen during moderate wind speeds – around six metres per second (that's a wind speed of thirteen miles per hour, which we might describe as a moderate breeze). This is characteristic of the contribution of an elevated source of emissions to air, so I would be off looking for a chimney to the north-west of the monitoring station. It might perhaps be a commercial or institutional boiler, and maybe one which isn't operating very efficiently, resulting in the discharge of higher levels of oxides of nitrogen than would normally be expected. A quick shufti using a widely available online mapping resource suggests a number of educational institutions in this direction which might run a boiler responsible for the high measured concentrations.

One of the strengths of the OpenAir system is that I was able to rapidly check how the bivariate plot looks for data recorded in 2015, and for data recorded in the first four months of 2017. The source of NO_x was present in 2015, but there was no clear indication of a source in 2017. Also, I could quickly check whether there was any evidence for a source of sulphur dioxide or PM_{10} in a similar location. There wasn't – which would make me think that the source I'm looking for probably involves combustion of natural gas which does emit oxides of nitrogen but doesn't emit PM_{10} or sulphur dioxide. A more detailed investigation of this unknown source of pollution would probably require some door to door enquiries, which might not make me very popular, particularly as the source seems to have gone away according to the 2017 measurements. Let's not forget that air pollution data doesn't lie – or at least, it has no inherent bias. Measurements might be misinterpreted, or they might perhaps be favourably interpreted with a view to supporting some agenda or another, but the measured levels are what they are. If the measurements indicate that, for example, there's a source of air pollution somewhere to the northwest, or vehicle emissions are not declining as fast as they would if every vehicle manufacturer were scrupulously complying with the emission limits, then that's probably a voice that we should listen to.

So far, we've been looking at measurements of oxides of nitrogen using chemiluminescence analysers: a very well-established technique going back forty years. How about the other air pollutants?

MEASURING SULPHUR DIOXIDE

In the UK, we have been measuring sulphur dioxide since the 1950s. This was a critical part of understanding the impact of exposure to sulphur dioxide during the smogs of the 1950s. For many years, the workhorse of the UK's air quality monitoring systems was something called the smoke and sulphur dioxide network. Sulphur dioxide was measured by bubbling air through a solution of hydrogen peroxide. The sulphur dioxide reacted with the hydrogen peroxide to form sulphuric acid in the solution. In fact, any acid present in the air sample would give rise to acid in solution, but the majority of acid in the air was due to sulphur dioxide, so it was a reasonably good indicator of airborne sulphur dioxide levels. The amount of acid was then determined in a laboratory, to give a reasonable measurement of the level of sulphur dioxide in the atmosphere. An instrument was developed which enabled eight 24-hour samples to be taken, so that an instrument could be left in the field for a week between visits to collect samples, replace the hydrogen peroxide solution and set it going again for another week. Adding a filter upstream of the bubbler liquid enabled the amount of smoke in the air sample to be measured as well. This method was cheap, cheerful and robust, and kept going from the 1950s until 2005, with over 2,000 sites listed in the UK air quality archive.

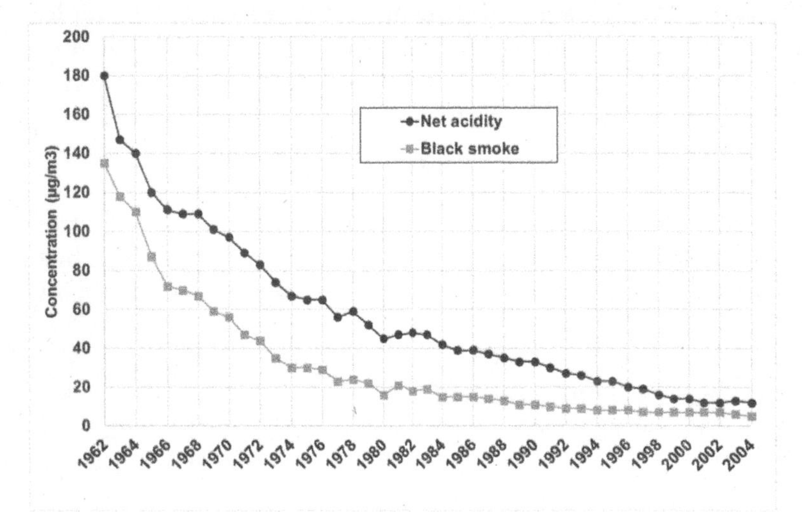

Annual mean smoke and net acidity concentrations from 1962 onwards[117]

The figure shows how measured acidity levels dropped by 93%, and smoke levels by 96%, between 1962 and 2004. The network was discontinued after the 2004–2005 monitoring year: its work was done, and there were more important measurements to be taken.

Nowadays, the standard method for measuring sulphur dioxide uses ultraviolet fluorescence from sulphur dioxide molecules. Sulphur dioxide absorbs ultraviolet light with a wavelength of 214 nanometres, so the first step is to shine UV light at this wavelength on an air sample. The sulphur dioxide then decays back down to a lower energy state, and emits ultraviolet light at a different wavelength – 350 nanometres. A detector placed at right angles to the original light source is used to measure the fluorescent UV light emitted by sulphur dioxide in the air sample. The intensity of this light is proportional to the concentration of sulphur dioxide. It's similar in some ways to the nitrogen dioxide analyser, but a simpler instrument in principle, because the fluorescence is initiated with a light source rather than by reaction with another chemical.

MEASURING HYDROGEN SULPHIDE

One adaptation of the sulphur dioxide analyser which I have found useful is a modification which enables it to measure hydrogen sulphide. Hydrogen sulphide isn't normally a big deal, so there isn't a national monitoring programme for this substance. However, it is one of the most smelly chemicals you're ever likely to meet. An instrument which can reliably measure even very low levels of hydrogen sulphide which are still enough to result in a strong smell can be a very useful tool to identify and deal with odour problems. The modification to the sulphur dioxide analyser is to add a catalyst to oxidise any hydrogen sulphide present to sulphur dioxide.

$$2H_2S + 3O_2 \rightarrow 2SO_2 + 2H_2O$$

The instrument can then be operated alternately to measure first sulphur dioxide, and then hydrogen sulphide (newly oxidised to sulphur dioxide) and sulphur dioxide already present in the sample. Subtracting one from the other gives a reasonable measurement of the hydrogen sulphide concentration. Hydrogen sulphide has a characteristic odour of rotten eggs – very distinctive, and it can occur where there are odour problems caused by sewage treatment or other industrial processes, particularly where waste water treatment processes are not working properly. Continuous measurement of hydrogen sulphide can be useful in understanding the characteristics of this kind of problem, and quantifying the odour problems which might result.

Handheld instruments using a gold film sensor can also be used for measuring hydrogen sulphide and other organic sulphide compounds at very low concentrations – again, low enough to be useful in investigating odour problems. These are flexible enough to be used to investigate odours around a possible source, and, with a bit of tender loving care, can also be plugged in and left to collect data.

MEASURING CARBON MONOXIDE

Carbon monoxide is usually measured using an infrared absorption technique, although there is not too much investment in monitoring carbon monoxide in the outdoor environment these days. Levels are so low that nowhere in the UK at least is at risk of exceeding the air quality standard. Doesn't mean you can ditch your household carbon monoxide analyser though: a build-up of carbon monoxide in the home is as dangerous today as it ever was, so keep checking those batteries!

MEASURING INORGANIC SUBSTANCES

There are two further groups of substances to think about – inorganic compounds, and volatile organic compounds. Inorganic compounds are substances like metals and soot which are solids at typical atmospheric temperatures. We typically measure airborne levels of these by drawing an air sample through a filter. The solid particles are retained on the filter, while the air and any other gaseous compounds pass through without stopping. The filter is then analysed to determine the amount of metals, or soot, or whatever. As long as you know how much air passed through the filter, it's pretty straightforward to work out the concentration of inorganic substances in the air sample.

MEASURING ORGANIC SUBSTANCES

Many organic compounds are present as gases in the atmosphere, and these are usually measured by capturing the substances on a material with a very high surface area. With the right combination of material and pollutant, the airborne chemical in question can very readily be captured on the surface of the solid material, a process known as "adsorption". These high surface area materials often look like small crystals or a powder – silica gel (sometimes provided as a small package in new luggage and similar products to absorb water vapour) is one material which is sometimes used for this purpose. Air is pumped through a tube containing the adsorbent material, and the material can then be analysed in a laboratory to identify the chemicals that are present, and their concentrations. That's the way we would carry out a site-specific environmental survey, but there is also the fag end of a national monitoring network which provides hourly measurements of some organic compounds. In the 1990s, there were thirteen hydrocarbon monitoring stations across the UK, but now only four stations are left – two in London, and two rural sites. They are still chugging away, measuring twenty-nine separate organic chemicals every hour. These monitors are there to fulfil our obligations to measure the concentrations of ozone precursors under the European Air Quality Framework Directive. They collect a lot of data, but I've never come across anyone actually doing anything with the numbers, other than just reporting it to the European Commission. Maybe one day it'll be useful to someone.

A small number of very low volatility organic compounds are measured using a combination of both the filtration and adsorption techniques, to make sure that both solid phase and vapour phase materials are captured. These include the very unpleasant dioxins and furans (which we looked at in the context of waste incineration in Chapter 8), and another group of chemicals which is almost as nasty: polycyclic aromatic hydrocarbons.

MEASURING PARTICULATE MATTER

And that pretty much leaves airborne particulate matter. In the past, the majority of airborne particulate matter was due to smoke from combustion of coal and oil. As well as commercial processes, industrial processes and power stations, coal burning in households was very prevalent in the UK for many years. Coal burning is now limited to a few properties and areas, particularly in parts of the country where coal mining used to be widespread. Some former employees in these areas have access to subsidised coal for use in the home, which provides a strong incentive to continue using solid fuels. Because of the prevalence of coal burning in times gone by, most airborne particulate matter was black in colour. This meant that you could measure particulate matter by taking a sample onto a filter paper and measuring how much the filter paper is discoloured. This type of measurement is carried out by shining a light onto the filter and measuring how much of the light is reflected back. This was the method used to analyse filters taken from the smoke and sulphur dioxide network.

That was fine for twenty or thirty years, until we inconveniently stopped burning coal in our homes. As we enthusiastically started to heat electric, and adopted the cook, cook, cookability of gas, airborne particles stopped being predominantly black, so you could no longer measure the amount of particles just by looking at colour changes. We had to start measuring the weight of particles more directly.

These days, two systems are pretty widely used for measuring airborne particulate matter. The first step for both methods is to use a sampler which selects the particle size fraction of interest. This is basically a bent pipe. As the air goes round the corner in the pipe, particles which are larger than, say, PM_{10}, can't get round the bend, and are taken out of the sampled air flow. Make the bend a bit tighter or the pipe a bit narrower and the instrument will select $PM_{2.5}$ or PM_1.

The most widely used monitoring system is called a Tapered Element Oscillating Microbalance, reinforcing any perceptions you might have that air quality scientists don't have a great way with words. Fortunately, it is universally known as a TEOM. The instrument works by passing the air sample through a filter, on which any particles are deposited. The filter sits on a little support which vibrates at a frequency which can be related to the mass of the filter, including all the material deposited on the filter. All you need to do is measure the vibration frequency, and the instrument converts this directly into a measurement of the extra mass on the filter.

That was fine when the TEOM was introduced, but it soon became apparent that the instrument was under-reading compared to the standard reference method, and the reason for this was – it was too hot. Unlike the reference method (which is a non-automatic system based on weighing filters), the TEOM requires the air sample to be heated to fifty degrees centigrade. This means that semi-volatile particles disappear from the sample and are not counted towards the measured level of particulate matter. We really need the total quantity of particles, because we're not usually breathing air at fifty degrees centigrade, so an amended system was introduced to give a separate measurement of volatile and non-volatile particulates. This amended system is called "Filter Dynamics Measurement System". Rather than removing water by heating the air sample, water is removed through a selective membrane in a dryer unit. So this system for measuring PM_{10} and $PM_{2.5}$ now has the snappy title of TEOM FDMS.

Going back to the mid-1990s, I had the privilege of installing and using one of the first of the alternative particulate monitors to be deployed in the UK. I seem to recall that it wasn't my decision to be an early adopter of a whole new method of measuring PM_{10} called a Beta Attenuation Monitor, but I was very much in the frame for trying to make the thing work. With any new technology, the first systems to be put on the market can be expensive and prone to unforeseen problems. As the technology develops and experience with the systems grows, costs come down and reliability improves. In the early days, however, we were still learning about the practical issues of air sampling, so trying to get an early BAM to work gave me many happy hours in the back of a cramped trailer in a small field next to the M60 motorway in north Manchester.

The BAM works by measuring the absorption of beta-particles by particulate matter on a filter. The greater the absorption, the higher the concentration of particles. Pretty simple in principle, and it works in practice, so I did come to like the BAM in the end. I liked it a lot more than the NO_x analyser we were using, even though the NOx box was a lot more established technology. Our instrument included a dust filter which I had to replace every week, which was conveniently situated on the front panel of the instrument. Unfortunately, the manufacturers hadn't worked out that this meant that the filter paper was vertical, and would invariably drop out the first sixty or so times that you tried to screw the cover back on. The biggest problem was self-inflicted, however: getting to the site and realising that I had left the trailer key or a vital screwdriver back in the office on the other side of Manchester. Still, I only did that five or six times, so it's not as though I failed to learn from my experiences...

Most monitoring for PM_{10} and $PM_{2.5}$ in the UK national air monitoring network is carried out using TEOM FDMS instruments and Beta Attenuation Monitors. These systems provide a measurement of the mass of particles in the PM_{10} and $PM_{2.5}$ size fractions. We need to follow this approach to end up with measured concentrations of PM_{10} and $PM_{2.5}$ as micrograms per cubic metre – that's a mass per unit volume of the air. However, there is evidence to suggest that the health effects of airborne particulate matter are linked to the *number* of particles we breathe in, rather than (or as well as) the *mass* of particles. You might expect the number of particles to be closely related to the mass of particles, but this isn't necessarily the case. One particle with a diameter of ten microns weighs as much as a thousand particles with a diameter of one micron. So a measured concentration of PM_{10} could be made up of a few big particles with diameters of about ten microns, or thousands of particles with diameters of about one micron –

or millions of particles with diameters of about 0.1 microns. Or, more likely, some combination of particles of different sizes. If the health effects of particles are linked closely to particle numbers, we should probably be measuring particle numbers – but at the moment, we just don't have enough information to tell us whether the health effects of particles are linked to the mass of particles, or the number of particles, or some combination. We do know that there is a link to the mass of particles, and there are standards for particle mass, so that's where our policy and monitoring effort is currently focused. A small network of instruments collects data on particle numbers at sites in London and Birmingham, and at two rural locations – maybe one day this network's time will come.

For specific site assessments, other techniques are available. One filter-based system gives a 24-hour mean concentration of particulate matter, and can be left in the field for up to fourteen days. It's not rocket science, but it's simple and reliable. Some instruments use light scattering techniques to give an indicative measurement of particulate matter in different size fractions. Light scattering methods can be very useful for a quick look-see kind of study, but the measurements are not to the same standards as measurements made using the TEOM or BAM techniques. I never quite trusted the light scattering instrumentation since putting one of them just outside our office in a cloud of smoke, to see it record essentially zero levels of every size fraction – total particulate matter, PM_{10}, $PM_{2.5}$, and PM_1. Now, it might have been down to me not setting up the instrument right, but as everything seemed to be operating normally, and I could see the smoke particles drifting across the grass outside the window, it did give me some doubts. But that's just me, and more rigorous verification studies do show good correlation with the gravimetric (i.e. weight-measuring) methods.

REMOTE SENSING BY SATELLITE

Nowadays, information on air pollution levels is becoming available from satellite-borne monitoring instruments. This exciting new data source gives valuable new information covering wide areas where fixed monitors have not previously been available. The figure below shows data from NASA's "Air Quality Observations From Space" website.

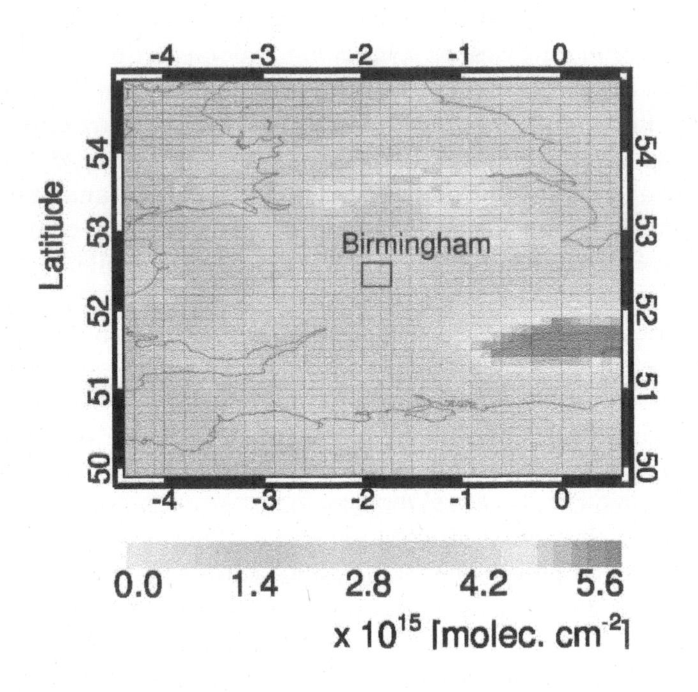

Satellite measurements of nitrogen dioxide in England and Wales, 2016[118]

These measurements do give a valuable perspective on wide-scale levels and trends in air pollution concentrations. However, there are a number of disadvantages. The data is only available at quite a coarse resolution of several hundred metres, or a few kilometres, whereas we are often concerned with changes in pollutant levels over a few metres. Also, the measurements represent the total amount of each pollutant in the column of air between the satellite and the ground, so they do not necessarily represent the concentration that people at ground level are exposed to. And finally, I'm pretty sure that this data will have no traction in influencing air quality policy and improvements until scientists stop using units like "$\times$ *10^{15} [molec. cm^{-2}]*" to give us a fighting chance of using plain English to explain what all this amazing data actually means. The nitrogen dioxide levels are presented as the total number of molecules in the troposphere above each square centimetre of the Earth's surface. How does that relate to the measurements that actually matter: that is, the amount of nitrogen dioxide in each cubic metre of air that we might breathe in? The way that satellite data is collected and presented remains pretty arcane, and I would hazard a guess that these kind of numbers don't mean much to most people working as environmental scientists, never mind the politicians and policymakers who have to actually decide on what action to take in response to air pollution. Where it does become more useful is looking at trends – identifying whether things are getting better or worse, and correlating these trends in the amount of nitrogen dioxide in the atmosphere to trends in emissions. For example, we can see improvements in nitrogen dioxide in the troposphere between 2005 and 2014 in cities across Europe, with improvements particularly apparent across northern Italy, the Madrid region, the Netherlands and south-east England. Meanwhile, things got a lot worse in Bangladesh, Pakistan, Angola, and the Middle East. So, notwithstanding my minor quibbles about molecules per square centimetre, watch out for more, and better, satellite data coming through the media in the coming years.

WHAT CAN WE DO ABOUT AIR POLLUTION?

So that's how to get hold of information on levels of air pollution: we can measure what's there now, and we can use models to fill in the gaps between measurement locations, and to predict what would happen in the future. So far, so passive: it's now time to start doing something about air quality. For a long time, right up to the introduction of air quality management in the 1990s, there was no direct connection between controls on emissions of air pollutants, and the effect that these controls would have on air quality. We haven't been short of controls on what goes up into the atmosphere – as far back as 1306, King Edward I proclaimed a ban on burning sea coal in London (yes, the same stuff that John Evelyn was still banging on about in 1661: it looks like frustratingly slow progress in improving air quality has been with us for over 800 years). As the industrial revolution took hold, the Alkali Act of 1863 was the first legislation which set controls on the air pollution that could be emitted by industrial processes – and backed it up with the forerunner to today's Environment Agency: the Alkali Inspectorate. Over the following century and a half, this legislative programme developed to provide for an integrated approach to managing pollution from industrial processes not only to air, but also to land, water and waste disposal. After the great smog of 1952, the Clean Air Act introduced similar controls to a wider range of mainly smaller scale processes. As we've already seen, controls on emissions from road vehicles began in the 1970s, and have continued to tighten, with the occasional blip from crafty and unprincipled vehicle engineers. And there was a third strand to controls on air pollution, through the land-use planning system. After the Second World War, the Town and Country Planning Act introduced controls on rebuilding and development in the UK through the production of local plans by local councils. As time went on, environmental controls became increasingly important components of the planning system, to the extent that air quality, among other environmental aspects, is an important component of many planning decisions. Very occasionally, the air quality impact of a new development is sufficient to lead to a refusal of planning permission. This doesn't happen often, but that's not necessarily because the system isn't protecting air quality: it's more to do with developers making sure that their applications comply with the national and local requirements on air quality before they get too far.

So we've got controls on industrial emissions, small-scale commercial and domestic emissions, vehicle emissions and new developments. Sounds good, but would this be enough to deal with air pollution and achieve air quality standards? Well, until we got going with local air quality management, nobody really knew if these controls would do a job on air pollution for us. As time went on, however, it became clear that piecemeal controls on different sources of emissions weren't going to be good enough, and so a programme of air quality management at a local level was introduced across the UK. For the first time, we took a step back and tried to understand the sources of air pollution, what effects they have on air quality, and what the future holds if we don't do anything more to improve air quality. That process used information on emissions to the atmosphere from all different kinds of sources, combined with measured levels of air pollution. We used dispersion models to fill in the gaps, although quite often there was no need to go to all the effort of a detailed and expensive modelling study – much simpler systems are available which we can use to check if there's a possible problem. We can then focus our attention on the problem areas, of which there are plenty.

As well as joining up the dots between air quality monitoring stations, computer models allow us to investigate and design measures which could be taken to improve air quality. This is the critical step: we want to do enough, but in a system driven by compliance with air quality standards, local councils don't usually have the inclination or indeed the authority to go beyond the bare minimum of just-about-complying. So, in areas where there is a problem, what's the best way to achieve the air quality standards? Well, every area has its own priorities. In a few places where sulphur dioxide was still above the standards when local air quality management took off, for example, the focus was on improving emissions from local industrial sources through the permitting process. Road traffic through a congested town centre is a more common air quality problem in the UK. Here, the focus might be on road improvements to speed up traffic flows, or maybe investments in schemes such as Park & Ride to shift traffic away from the centre of town. And where an air quality problem is caused by large volumes of traffic on trunk roads… well, to be honest, we've not done a whole lot about that, apart from wait for improvements in emissions technologies to work their way through the vehicle fleet. For example, the ten local authorities in Greater Manchester work together to evaluate and improve air quality throughout the city. Their annual update highlights the importance of road traffic for air quality in Manchester, with particular mention of the M60 motorway which runs through the city for a distance of fifty-eight kilometres, in some areas passing close to schools and houses. The report outlines sixty-nine separate initiatives to improve air quality, of which just one focuses on the M60 motorway. This has the laudable aim of reducing congestion on the motorway, but the lamentably unambitious target of being "air quality neutral". So it looks like Manchester, a classic case of a city where trunk road traffic needs to be addressed to make a real dent in air pollution levels, is doing everything except addressing trunk road traffic. The trouble is that main roads are the responsibility of a national agency: local authorities don't have any say in dealing with motorway traffic. And who wants to see restrictions on traffic travelling from one side of the country to another, just to improve air quality at a few houses in Manchester? Apart from the people living in those houses, of course.

At least the air quality monitoring and modelling studies for Manchester and elsewhere tell us what the problem is, even if we haven't yet bitten the bullet sufficiently to do much about it. Using air quality modelling techniques, we can forecast what the effect of measures such as the "Manchester Sixty-Nine" will be on air quality. Alongside that, we can estimate the cost of each of these measures, and work out the most cost-effective means of achieving our objectives – whether that is to comply with an air quality standard, or to make a dent in the 30,000 or so early deaths caused by air pollution every year. The science is doable. A lot of the measures are affordable, and might even be popular, and might even bring some additional benefits alongside air quality improvements. But will we do them? What does the future hold?

CHAPTER 10

THE FUTURE: MY LIFE, MY CAR, MY TOWN, MY WORLD

And Back Around the World to Forty Thousand Kilometres

THE LONG VIEW

What does the future hold for the atmosphere? Will there still be air for our children and their children to breathe in thirty years' time? Or 500 years' time? Will we be concerned about the global climate or the ozone layer? Will air pollution still be bringing an early end to millions of lives every year, or will we be more concerned about the occasional smell? Whatever happens to the atmosphere, I'm pretty sure twenty-sixth century man and woman will still be worrying about whether their house is worth more than the Joneses across the other side of town.

At the global scale, there should still be plenty of oxygen around, so that's a good start. In 500 years, at the present trend, oxygen levels in the atmosphere might drop from 21% to 20%. And that's assuming that we continue to consume oxygen at the current rates, with the same net balance between billions of breathers in the human population, use of oxygen in combustion processes, and production of oxygen from photosynthesis. We should be able to breathe pretty well in an atmosphere containing 20% oxygen, and if things start getting desperate, growing more plants and burning less fuel would start to restore the balance back up towards 21%.

If oxygen levels do drop off a bit, it might affect a twenty-sixth century Usain Bolt's chances of breaking the 100 metres record in the 2520 Olympics, so keep an eye out for that.

That's assuming business as usual. But to be honest, who knows what we'll have done to the global climate in the future? Over a timescale of hundreds of years, it's likely that we'll use up pretty much all the accessible fossil fuel reserves, so we'll be relying on other sources of energy. Once we've done that, and once the long-lived greenhouse gases that we've put into the atmosphere have finally been removed, you might hope that the global climate would be on the way to recovery. Unfortunately, it looks like that isn't the case: the Intergovernmental Panel on Climate Change[119] comments that: "*Most aspects of climate change will persist for many centuries even if emissions of CO_2 are stopped. This represents a substantial multi-century climate change commitment created by past, present and future emissions of CO_2.*" So it looks like we're stuck with an altered climate, and all that comes with that in terms of higher sea levels and unpredictable weather patterns for example, for a very long time to come.

How we manage the long-term transition away from fossil fuels is going to be one of the determining factors of quality of life over the next couple of hundred years. Maybe we will move backwards to reliance on biomass like we did a thousand years ago, or maybe we'll progress forwards with existing and maybe new renewable energy technologies. As the availability of fossil fuels decreases in future decades and centuries, sooner or later that's going to result in a reduction in the use of these fuels, whether we like it or not. It might be scant consolation as energy prices and global tensions increase, and we struggle to provide heat, power, water, food and mobility, but the end of fossil fuels will at least be good news for air pollution. Well, it will if we've developed new energy efficient and low emissions technologies and lifestyles. If alternatively we just go back to burning wood and animal dung for cooking and heating, we may well continue to suffer from the debilitating effects of air pollution. However, if it does come to that, the human race will probably have more urgent problems to worry about.

STILL GETTING TO GRIPS WITH URBAN AIR QUALITY

At the urban scale, improving air quality in the UK will be focused on dealing with relatively small-scale problems in urban centres. The latest assessment from the government shows that nitrogen dioxide continues to be problematic in city centres throughout England and Scotland.[120] It's a limited problem – it's not as if nitrogen dioxide is above the standard throughout, say, Birmingham: the areas where levels are too high tend to be close to main roads, particularly where these are congested and the buildings are continuous along the roadside. As well as bringing people closer to the vehicles which emit the pollution, a continuous row of buildings on each side of a road makes it harder for pollution to escape from the road and disperse over a wider area. Slightly romantically, we call this a street canyon, reflecting the vertical sides of the buildings and their influence on pollution, rather than any resemblance to the Colorado River in Arizona. But in an urban setting, a street canyon can result in substantially higher levels of air pollutants than you'd otherwise expect. This has resulted in some small-scale but surprisingly severe air pollution problems in small towns – for example, the genteel Oxfordshire communities of Henley and Wallingford,[121] which are not otherwise widely known for their airpocalyptic levels of pollution. We're now doing a lot of work to identify the best measures for improving air quality in these city centres as quickly as possible. But we're not using the established process of designating Air Quality Management Areas: instead, a series of Clean Air Zones is being designed and implemented in city centres throughout the UK. These focus on traffic emissions, with options for charging and non-charging zones, and maybe metaphorically bypassing AQMAs, which have been around for a decade and a half without fully dealing with air pollution problems, is a good idea. The new CAZs can perhaps take advantage of renewed public interest in air quality to bring lasting improvements in areas where AQMAs haven't really delivered. So we now have CAZs alongside our AQMAs, which will hopefully deliver VG air quality PDQ.

In fact, existing policies are expected to deliver compliance in almost all the zones, with the exception of London, and the latest round of air quality plans, Clean Air Zones and all, are designed to make sure that this happens as early as possible, as we are required to do under European legislation, and the national legislation which derives from it. Which slightly begs the question of what happens if, or I suppose I should say when, we in the UK leave the European Union. While we can expect to keep the same air quality standards for the foreseeable future, they will no longer come loaded with the threat of action by the European Commission if we don't comply with the directive requirements. The air quality standards may still have teeth, and can still be backed up by the British courts, but will they be taken as seriously when they are just a domestic matter? My hope is that this might become irrelevant, as we move into an era which is driven by a popular groundswell demanding action to minimise the huge health burdens of air pollution by reducing pollution everywhere, rather than action in ever-diminishing areas to comply with air quality standards which were set on the basis of out-of-date science. Now we have a good idea of the long-term health burden of air pollution, we have the tools in our hands to view and budget for the management of air quality on the basis of improving everyone's health, rather than just tidying up round the edges to make sure we comply with air quality standards. There's still a role for air quality standards as minimum thresholds to make sure that no one experiences excessive air pollution, but we shouldn't just be looking to get over the finishing line of achieving a standard and calling it a day there. And how much more is that the case in parts of the world where levels of air pollutants don't even come close to meeting air quality standards.

TRANSPORT EMISSIONS

Although we expect to achieve the air quality standards everywhere except London by 2027, experience of the past twenty years of air quality management suggests that we can't be 100% sure that this will really happen. As we saw in Chapter 6, one of the main areas of uncertainty in these forecasts is in the emissions which occur from vehicles as they are driven on the road, compared to the emissions that are measured when vehicles are tested, for both legit and non-legit reasons. I suppose there is a silver lining of sorts: looking to the future, there is still much to be gained from reducing vehicle emissions to match the existing standards in the real world, and then going on to deliver ongoing improvements over the coming decades. This is one reason why we expect the air quality standard for nitrogen dioxide to be achieved almost everywhere in the UK by 2027.

Some of these vehicle emissions improvements will go hand-in-hand with reducing greenhouse gas emissions. More energy efficient vehicles will (probably) emit lower levels of all pollutants, both air pollutants and greenhouse gases. Other improvements might require a trade-off to be made between reducing greenhouse gas emissions, and reducing air pollutants, just like we did with catalytic convertors three decades ago. The recent commitment by Volvo to fit electric engines to all its vehicles, either as hybrids or as pure electric vehicles, perhaps signals the future for urban traffic in the medium term. Of course, electric vehicles require an electricity supply, but electric power is definitely a big step in the right direction to improve air quality in urban hot-spots. To take the final steps in eliminating pollution from urban traffic will require fundamental changes in our approach to mobility – do we need to move people around cities? And to the extent that we need or aspire to mobility, can we find ways of eliminating the private car as a means of getting people from A to B?

As we've seen, the major source of health impacts due to air pollution in the UK, and worldwide, is fine particulate matter. In the UK, we've roughly halved emissions to air of fine particulates since 1992, and that's been matched by a reduction in measured airborne concentrations over that time. So we've probably also reduced the number of early deaths resulting from exposure to air pollution by about 30,000 per year over twenty-five years. That must be good news! But can we do better? Most forecasts don't anticipate significant further reductions in emissions of $PM_{2.5}$ in the coming years. That seems surprising at face value, because there are opportunities to deliver further improvements to $PM_{2.5}$, which would bring further reductions in premature deaths. But the policy drivers for improving air quality in the UK at least are based on compliance with air quality standards and guidelines. In the case of $PM_{2.5}$, we have already achieved the air quality standards pretty much everywhere. Without a strong driver for further improvements in $PM_{2.5}$, it's hard to see the required steps being taken. Of course, saving tens of thousands of lives a year ought to be a strong enough driver for action.

VALUING THE BENEFITS

We could at least make some estimates of the costs of reducing $PM_{2.5}$ emissions, and the number of lives that would be saved. We could then look at other interventions which might be considered to save lives – for example, new medical treatments, improvements in road safety, or raising public awareness about diet and lifestyle. This would enable a rational decision to be taken about whether it's worth spending hard-earned money on further reductions in $PM_{2.5}$ emissions.

The trouble is finding the framework for doing this kind of evaluation. We can calculate the cost of interventions to improve air quality, such as changing fuels, introducing emissions charging schemes, or public information campaigns ("Attention please! Do not install a wood fired stove in your house!") We could estimate the benefits of interventions in reducing $PM_{2.5}$ emissions and thereby reducing public exposure to $PM_{2.5}$. We could go further and calculate the economic value of these benefits. These benefits and their associated economic value would be real, but would not be attributable. You couldn't point to specific deaths which have been avoided by improving or extending controls on $PM_{2.5}$. So it would be impossible to identify where the cost savings have accrued on an individual basis. However, air quality modelling studies would enable us to identify the geographical areas where benefits might accrue. The last piece in the jigsaw for rational decision-making is not scientific, it's a matter of accountancy: the costs of improving air quality would fall on stakeholders such as local authorities, businesses, motorists, consumers, residents. In contrast, the benefits would be gathered by the health and emergency services as they don't have to treat so many people suffering from respiratory and cardiovascular diseases. Employers and businesses would benefit from a healthier and more productive workforce. The benefits would be most directly felt by the individuals who avoid pneumonia and heart attacks thanks to improved air quality, but of course, they won't know who they are either. With the costs and benefits accruing to different, not very well defined stakeholders, it's hard to make the financial case for investment in air quality improvements.

One initiative has been introduced to encourage improvements in air quality to be considered at a strategic level, and that is the little-known "Public Health Outcomes Framework". This closely-guarded secret was introduced when local authorities in England were handed responsibility for public health in 2013, and it was updated in 2016. A total of sixty-six targets are set for public health outcomes against which local authorities can be measured and benchmarked. One of these indicators relates to air pollution: *"Public Health Indicator 3.01 Fraction of mortality attributable to particulate air pollution."*[122] Local authorities are seeking to improve their performance in relation to all these indicators, of which number 3.01 is, of course, just one among sixty-five others. Leaving aside the large number of indicators that a public health authority is responsible for, this approach looks like a great means of delivering genuine and measurable public health benefits due to air quality improvements, and quantifying these at an appropriate level of detail.

But there is a problem, and it's to do with how the indicator operates. This indicator is based on calculated levels of $PM_{2.5}$ across a local authority's area. These levels are determined from a combination of air quality monitoring data and a modelling analysis based on the national atmospheric emission inventory. The trouble is that there is a significant disconnection between actions that a local authority could take to reduce $PM_{2.5}$ levels, and that working through to reductions in modelled $PM_{2.5}$ concentrations, and hence a reduction in the fraction of mortality due to air pollution. For example, a local authority could carry out a campaign to reduce particulate emissions from building sites, but if there is no air quality monitoring near these building sites (there probably isn't), and if the compilers of the national inventory are not able to make changes to the inventory to reflect this kind of local initiative (they definitely aren't), then it would not come through in the figures. And many of the measures that might improve levels of $PM_{2.5}$ are in any case outside the powers of local authorities... and the indicators do not have binding targets associated with them. Indeed, the Public Health Outcomes Framework is not used for performance management, it's just there so that local authorities can benchmark their performance against other authorities. In that case, it seems like a lot of effort, and dare I suggest, a missed opportunity to make the most of the wide range of data gathered under the Framework. Even so, it's a start, and maybe if air quality continues its rise up the news and political agenda, further action to improve air quality will become not just an aspiration, not just a target, not just sensible, but unavoidable. Public health indicator 3.01 could then be adapted and strengthened to form the basis of a final push, not just to achieve some notional standard, but to genuinely minimise $PM_{2.5}$ – get it as low as possible.

AIR QUALITY AROUND THE WORLD

At a global level, however, the picture is a bit different. Levels of airborne particulates throughout much of the world, and particularly in large cities, are much higher than is typical in our windy corner of north-west Europe. Back in Chapter 3, we saw that average $PM_{2.5}$ levels are above 50 $\mu g/m^3$ in seventeen countries which account for almost half the world's population (four-fifths of these people are in India and China). While this is bad news for people living and working in polluted cities, it does mean that the opportunities are there to deliver substantial reductions in public exposure to airborne particulate matter. By doing that, we can improve respiratory health and avoid millions of unnecessary premature deaths. There are legitimate questions about who pays for the changes that are needed to deliver these improvements, and where this kind of investment should sit in the range of other demands placed on cities worldwide. But there are also good answers, firstly and most importantly in the genuine and quantifiable health benefits that result from improvements in air quality. As I mentioned previously, we can't necessarily identify individuals benefiting from improvements in air quality, but we can quantify these benefits at a community-wide level. And a lot of the steps needed to improve air quality where it is poor are not rocket science, and nor are they expensive. Simple steps can make a big difference – perhaps a matter of surfacing some roads, doing better with controls on construction dust, or providing people with alternatives to coal and wood for heating and cooking. Such steps might well be practicable at minimal cost in many parts of the world. And of course, once you've plucked the lowest hanging fruit by taking the cheapest and easiest steps, there are plenty of more expensive and harder to implement options for improving air quality to consider. However, taking even the simplest of these steps requires commitment and support from politicians, regulators and community leaders. In all too many places, this doesn't happen. Vested interests, other priorities, a lack of expertise or individual failings can all contribute to making air pollution too hard or too low a priority to be dealt with.

What gives me the most hope that real change is on the way, or even under way, is how well informed and angry people are about air pollution. Awareness of, and interest in, air quality has never been higher. I've mentioned the BBC's *So I Can Breathe* and *The Times Clean Air For All* campaigns. Stories with an air pollution angle make the TV and radio news every few days nowadays, finding a place between detailed analyses of Brexit and Strictly. *Woman's Hour* on BBC Radio 4 recently ran an item, asking what parents can do to protect their children from air pollution. And it's not just Radio 4 listeners – as I mentioned in Chapter 6, British tabloid newspaper *The Sun* published forty-four separate items on "air quality" in the first four months of 2017, some of them picking up on genuinely important issues to do with diesel scrappage and the impacts of air pollution on individual and population health. To be fair to *The Sun* and not wishing in any way to undermine its reputation in the pantheon of British newspapers, there were plenty of stories about B-list celebrities and the England football team as well.

But it's not just in the UK that interest in air pollution is at an all-time high. I've hinted a few times at the term "airpocalypse" which has become common currency in northern China to describe the combination of smoke and sulphur dioxide from burning coal with adverse weather conditions during the winter which results in the kind of choking smogs that we thought had disappeared along with the 1950s (you'll recall that the Body Shop has also used the term "airpocalypse" in its bus shelter campaign, whereas the use of the term in relation to air quality in Henley and Wallingford is all my own work). Air pollution in China is no longer the preserve of a few environmental campaigners. As in the UK, air quality in China has become front-page news – as I write, a quick look at the official Xinhua news agency reveals "*Beijing-Tianjin-Hebei region to suffer from ozone pollution,*" for example, and I'm sure there will be more recent examples by the time you read this. The pressure to improve air quality comes from members of the public, from industrialists and business people, from investors, from tourists – and also from the top, as ministers and even the president of China start to speak out on the topic of air pollution.

The dream and the reality of Beijing, December 2016[123]

This has forced politicians to face the complicated scientific, technical and societal issues associated with improving air quality in China. There is a commitment to spend serious money in tackling air pollution – the Asian Development Bank, for example, is putting two and a half billion (yes that's billion, not million) dollars into air quality improvements in the Beijing-Tianjin-Hebei region of north-east China (which also goes by the pleasing if mystifying acronym of jing-jin-ji) over a five year period. A story from Xinhua in February 2017 highlights a range of steps being taken to improve air quality in jing-jin-ji.[124] The report goes on to talk about failings in enforcement and regulation by some of the city authorities in this region, including naming and shaming those which have not delivered the goods. It's a mature approach to dealing with air pollution. China has moved on from seeking overseas aid to help with setting up the structures for dealing with air pollution (which was why I went to China back in 1995), and is now in the middle of designing and implementing controls. That's why sulphur dioxide emissions in China have started to turn the corner, and are at last on the way down, as we saw in Chapter 3.

Alongside industrial pollution controls, China has introduced controls on vehicle emissions, backed up with production of cleaner vehicles and provision of public transportation systems. For example, in Shenzhen (a city with a population of twelve million), all sixteen thousand buses and twenty-two thousand taxis are now electric, thereby effectively replacing the danger posed by air pollution emissions with the danger of creeping up on people from behind without making any noise. Nevertheless, air pollution in China is by no means under control yet: the World Health Organization estimates that average $PM_{2.5}$ levels in China are almost 60 $\mu g/m^3$ – six times the WHO's guideline. There remain substantial challenges for air quality in parts of China, which will require increasingly sophisticated analysis and control measures, and that's where attention is now beginning to focus. And that's why I expect air pollution in China to continue to fall over the coming decades, admittedly from a pretty astronomical starting point, so it will take many years to bring pollution close to a level that China's own population might consider acceptable.

Interestingly, India overtook China as the country with the most polluted air in the world in 2017, according to the State of Global Air report, produced by the Health Effects Institute and the Institute of Health Metrics and Evaluation.[125] In this context, "polluted" air is measured by the number of premature deaths caused by exposure to $PM_{2.5}$ – so to pick up the unwanted title of "most polluted air" you need a combination of high levels of $PM_{2.5}$ and a big population. And India undeniably has major air pollution problems, and of course a large and growing population. But in spite of this, there doesn't seem to be quite the same groundswell demanding improvements to air quality in India as we have seen in China. Maybe this will come – particularly if interventions in China continue to deliver ongoing improvements in air quality. People living in other cities in Asia and around the world will see the improvements delivered in China and want a piece of the action for themselves.

Meanwhile, in India at least, there has been very limited progress in improving air quality. Vehicle emission standards are in place, but lag a long way behind those used in Europe and elsewhere in Asia, and are not well enforced. Meanwhile, vehicle ownership in India has increased by about 12% per year, every year since 1951, a phenomenal rate of increase which shows no sign of slowing down.[126] With such prolonged and substantial growth in road traffic, it's not surprising that urban air quality in India is among the poorest in the world. India is also expanding its coal-fired electricity generation capacity. This may actually bring a net benefit for air quality, if the electricity generated is used, along with other cleaner fuels, to reduce the use of biomass (wood and animal dung) in domestic stoves – fumes from the use of chullah stoves were estimated to be responsible for almost half a million deaths in India in 2011. That's a big "if", however, and concerns about the expansion of coal-fired power stations and the effectiveness of industrial pollution regulation in India remain. On top of these ongoing sources of air pollution, every year the Diwali celebrations in many Indian cities feature candles, bonfires and firecrackers which give a huge and unwelcome boost to levels of PM_{10} and $PM_{2.5}$. In October 2016, for example, this resulted in a peak in $PM_{2.5}$ levels of 890 $\mu g/m^3$, as the residents of New

Delhi woke up the morning after the end of Diwali to stagnant air and thick smog.[127] There are calls for a rethink in how Diwali is celebrated in India, and that's important, but the bigger picture is the long-term impact of airborne particulate matter, estimated by the World Health Organization to be 66 $\mu g/m^3$ – that's six and a half times the WHO guideline.

With some investment, a dose of political will, and public support, substantial improvements could be made in the worst polluted cities of the world. The opportunity is there to save millions of lives and improve the quality of life for billions of people. Many of the most basic changes wouldn't require major financial support – for example, improved management of dust from building sites, or better enforcement of existing controls on vehicle and industrial emissions. Worth doing? I think so.

CLEAN CITIES

Once we have carried out some pretty basic steps to bring the worst cities in the world in line with their peers, making widespread further improvements in air quality is going to be a lot harder. It's going to require fundamental changes in how we provide mobility and energy in cities. We will need to find ways of improving air quality which align with wider societal changes and aspirations. Imposing regulatory-driven solutions to reduce air pollution is OK and a necessary step, but will only get us so far. To deliver real improvements, air quality improvements must be part of a bigger picture. Technologies or lifestyle changes which reduce the need for routine, daily travel have the potential for significant environmental improvements. Internet connectivity does, of course, offer the possibility for reducing the need to travel – I myself only travel the sixty miles to my office once a week, thereby saving a notional 480 vehicle miles each week. Thank goodness there's no decent public transport option, otherwise I'd only be saving 480 public transport miles… I work remotely firstly because I can, and secondly because I don't want to spend three hours a day in my car, with fuel and garage bills to look forward to when I finally get home. Certainly, working from home is not open to everyone but this may be an option for more people than the 14% of the UK workforce who currently work remotely. A fit-for-purpose public transport system can also make a real difference to a city which would otherwise suffer from road traffic on congested streets.

Some European and North American cities have managed to deliver low pollution levels using a combination of measures – efficient, reliable and affordable public transport; high levels of cycling; Internet connectivity; and maybe a cool, windy climate conducive to the dispersion of emissions and replacement with clean air. The cleanest cities in the world were recently reported to be:[128]

- Calgary, Alberta, Canada
- Ottawa, Ontario, Canada
- Helsinki, Finland
- Stockholm, Sweden
- Zurich, Switzerland

I've removed half the cities from the original list because they have populations well below a million (Whitehorse in Canada, Tallinn in Estonia, and three US cities: Santa Fe, New Mexico; Honolulu, Hawaii; and Great Falls, Montana). The methodology for determining the top ten cities was suspiciously opaque, but the full list does suggest that being quite a small place helps if you're looking for good air quality – from the original list, Whitehorse has a population of under 30,000 and Santa Fe and Great Falls aren't much bigger. On the other hand, Stockholm has a population of over two million, and Calgary, Ottawa, Helsinki and Zurich are all above a million, showing that it is possible for a reasonably large city to have good air quality. All these cities share a highly developed public transport system, and Stockholm and Helsinki also have a reputation of being welcoming to cyclists.

Some of the factors which deliver good air quality are not readily transferable – as well as being quite small, location by the sea helps, and being a long way from any other cities is also useful. And all five of the cities listed are in temperate areas, with annual average temperatures between four and nine degrees centigrade.[129] Not much that a city like New Delhi with its population of about twenty-five million, average temperature of twenty-five centigrade, and 1,250 kilometres to the sea can do about that. Other factors like strong urban planning and design are applicable anywhere, even though they will take a long time to deliver significant air quality benefits from a standing start. But it's hard to see cities which have grown organically without effective public transport systems, or a culture and infrastructure which enables cycling to be a viable option for commuters, making the fundamental changes that would be needed to improve air quality. What's more likely is some focus on cost-effective ways of improving air quality. In many of the most polluted cities in the world, there is much that can be done at low or even zero cost to improve air quality. Enforcing existing legislation is always a good start: often, there are limits on emissions from some of the most important sources of air pollution, power stations and road vehicles for example, but if no one measures or checks what is actually coming out of chimneys and exhaust pipes, it's no surprise if the emission limits aren't complied with in practice. Regulation might be ineffective due to under-funding, or a lack of competence and expertise on the part of the process operator or regulator, or worse: there may be corruption going on. Enforcement of emissions and planning controls provides many opportunities for backhanders to enable those with a vested interest in continuing to pollute to do so. Authorities might allow development that isn't appropriate, or accidentally fail to notice the use of polluting fuels, or schedule checks to avoid times when emissions are high. So it's important to get systems in place which will enforce and back up the controls on air pollution that are already in place. There's not much point in developing more sophisticated air quality management strategies if the basics are missing. As we in Europe and the United States found, when trying to understand why diesel-engined vehicles were emitting suspiciously high levels of oxides of nitrogen.

In the medium term, there's plenty more that could be done to improve air quality in the cities with the highest levels of air pollution – dealing with sources like the chullah stoves, coal burning in small-scale brick kilns, and also investing in bigger and better public transport networks. It all takes money and time, and agreement on who pays for the improvements will be critical for successful implementation.

Can we raise our sights any higher than bringing the cities with the poorest air quality up to the standards currently enjoyed elsewhere in the world? Yes, I think we can, but the radical changes needed to finally nail air pollution will have to wait. Wait for what? For the revolution? Well maybe yes: sooner or later there will be an energy revolution as fossil fuels start to run dry, and prices start to go up. And up. If we haven't made wholesale and fundamental changes in lifestyle and technology to accommodate the changing energy landscape by that time, then they will be forced on us. We've got a while to start making changes before our non-renewable energy runs out, although of course the longer we wait, the more people will be affected by the air pollution which inevitably goes with the use of fossil fuels. But we are going to need time to move to a post-fossil fuel world, with all its potential benefits and inevitability.

IMAGINE A WORLD…

Imagine a world where the norm is to live, work, shop, and spend our leisure time within walking distance of our homes. Not only that, but we won't be shifting food around the world: that will also be limited to what can be grown and made more or less on our own doorstep. It's not that long-distance transport will be impossible – we won't forget the technologies that currently move us around the place – it'll just be very expensive and you won't want to waste your money on flying mangetout from Kenya to your local branch of Tesco, not if there's a local turnip on offer. So I envision a world where our more limited resources of renewable fuels and electricity are used for connectivity rather than for mobility. Perhaps we'll end up with a curious mix of state-of-the-art low energy technologies, and a return to pre-industrial lifestyles. We'll be going places by bike or on foot. We won't take holidays on far-flung beaches, but instead we'll spend more time with our friends and families. I'm not sure what jobs we'll be doing, but I'm sure that there will continue to be highly technical work that needs to be done, alongside keeping the vegetable patch weeded. Sounds like I'm predicting *The Darling Buds of May* with Wi-Fi.

It feels like I had a bit of a premonition of a post-fossil fuel world as I walked up the Budi Gandaki Valley in the Nepal Himalaya. We only went a couple of hundred yards up the road from our first night's accommodation before any prospect of motorised transport came to an abrupt end. Transport into and out of the mountains was horse, donkey, yak or human powered. Higher up in the valley, as we approached the village of Samargaun, it was ploughing day, and it looked like the whole village was out in the fields with pairs of yak, busily ploughing to prepare for planting potatoes and barley, in a scene which can hardly have changed in centuries. And when it was lunchtime, while the older people chatted and rested, everyone under the age of twenty-five got their phones out and started playing Candy Crush Soda and Instagramming their friends. Maybe Nepal is way ahead of the rest of us.

That's all speculation. But however the future looks, once we get over our addiction to fossil fuels, it'll be better for air pollution. It's something to look forward to, but like any addiction, it's going to be tough to get ourselves off something that we've had a couple of hundred years to get used to. We are starting to make steps in the right direction with electric vehicles and renewable technologies. But doesn't it need to go further than that? To give up our aspirations for travel, mobility and goods which have fossil fuels embedded in them. I'm as addicted as anyone else, and I can't see much prospect of wholesale change towards a sustainable lifestyle until we're pushed towards it by the relentless forces of economics and the finite nature of fossil fuels.

Until then, there's still plenty to be done. We'll perhaps be able to improve our knowledge of planets beyond our solar system with habitable atmospheres, though as a viable solution to air quality problems, that's on the speculative side of bonkers. Now that we're taking air quality seriously around the world, I think we'll find ways of solving not just the global challenges, not just the city or local air quality problems, not just my own respiratory health difficulties, but all of them together – and delivering a fairer, more enjoyable and more sustainable quality of life at the same time.

Meanwhile, I'm passing the time by hoping that my family and I aren't among this year's seven million premature deaths which are caused by air pollution. Or next year's seven million. Or the year after that. Or the year after that. We've got a long way to go.

NOTES

1 Oxford Living Dictionaries, https://en.oxforddictionaries.com/definition/atmosphere
2 The Extrasolar Planets Encyclopaedia at http://exoplanet.eu/catalog/
3 Anastasi, C.; Broomfield, M.; Nielsen, O. J.; Pagsberg, P., "Kinetics and mechanisms of the reactions of CH2SH radicals with O_2, NO, and NO_2," J. Phys. Chem. (1992), vol.96, pp. 696–701
4 Interview with Sting carried out by Richard Cook in a December 1983 issue of New Musical Express magazine http://www.sting.com/news/article/76
5 Max Planck Institute, "Atmosphere around low-mass Super-Earth detected" https://www.mpia.de/news/science/2017-03-GJ1132b
6 Chassefiere, E.; Berthelier, J-J.; Leblanc, F.; Jambon, A.; Sabroux, J-C.; Korablev, O., (2005) "Venus atmosphere build-up and evolution: where did the oxygen go? May abiotic oxygen-rich atmospheres exist on extrasolar planets? Rationale for a Venus entry probe." Abstract available at: http://www.ims.demokritos.gr/IPPW-3/index_files/Book%20of%20abstracts.pdf
7 Carl Zimmer in the New York Times, 3 October 2013, "The Mystery of Earth's Oxygen", http://www.nytimes.com/2013/10/03/science/earths-oxygen-a-mystery-easy-to-take-for-granted.html
8 "The human physiological impact of global deoxygenation", The Journal of Physiological Sciences. 67 (1): 97–106 https://en.wikipedia.org/wiki/Atmosphere_of_Earth, citing Martin, Daniel; McKenna, Helen; Livina, Valerie (2016).

9 From Peter Ward (2006), Out of Thin Air: Dinosaurs, Birds, and Earth's Ancient Atmosphere. Washington, DC: Joseph Henry Press, p116

10 Scripps O2 Program, Global Oxygen Measurements, http://scrippso2.ucsd.edu/

11 Photograph: Simon Matthews, 2017

12 Encyclopaedia Britannica blog, 5 January 2012, http://blogs.britannica.com/2012/01/how-much-does-earth-atmosphere-weigh/

13 Fullekrug, M.; Takahashi, Y., "Sprites, Elves and their Global Activities," Journal of Atmospheric and Solar-Terrestrial Physics Volume 65, Issue 5 (March 2003), Pages EX1–EX8, 495–660

14 World Health Organization, "7 million premature deaths annually linked to air pollution," 25 March 2014, http://www.who.int/mediacentre/news/releases/2014/air-pollution/en/

15 World Health Organization, "Preventing disease through healthy environments: a global assessment of the burden of disease from environmental risks," March 2016, https://www.who.int/quantifying_ehimpacts/publications/preventing-disease/en/

16 World Health Organization, "Tobacco key facts," 9 March 2018, http://www.who.int/mediacentre/factsheets/fs339/en/

17 European Association for the Study of Obesity, "Obesity Facts and Figures," http://easo.org/education-portal/obesity-facts-figures/

18 World Health Organization, "Drinking-water key facts," 7 February 2018, http://www.who.int/mediacentre/factsheets/fs391/en/

19 World Health Organization, "Global Health Observatory Data," http://www.who.int/gho/road_safety/mortality/en/ (data for 2013)

20 Climate Depot, "On Global Warming, Follow the Money: U. S. Spent $165 Billion on climate change," 15 July 2014, http://www.climatedepot.com/2014/07/15/on-global-warming-follow-the-money/

21 Hansen, J.; Sato, M.; Kharecha, P.; Beerling, D.; Berner, R.; Masson-Delmotte, V.; Pagani, M.; Raymo,M.; Royer, D.L.; Zachos, J.C., "Target atmospheric CO2: Where should humanity aim?" Open Atmos. Sci. J. (2008), vol. 2, pp. 217–231

22 NASA, "Global Greenhouse Gas Emissions Data," https://www.epa.gov/ghgemissions/global-greenhouse-gas-emissions-data

23 United Nations Economic Commission for Europe, "Clean Air," https://www.unece.org/env/lrtap/welcome.html

24 Data taken from Department for Transport (December 2016), "Transport Statistics Great Britain 2016"; Department for Business, Energy and Industrial Strategy, "National Atmospheric Emissions Inventory," http://naei.beis.gov.uk/

25 Greater London Authority, "London Atmospheric Emissions Inventory 2013" https://data.london.gov.uk/dataset/london-atmospheric-emissions-inventory-2013
26 Peter Edwardson, SpeedLIMIT Petrol Prices, http://www.speedlimit.org.uk/petrolprices.html
27 Transport and Environmental Analysis Group, Centre for Transport Studies, Imperial College London (April 2013), "An evaluation of the estimated impacts on vehicle emissions of a 20mph speed restriction in central London," https://www.cityoflondon.gov.uk/business/environmental-health/environmental-protection/air-quality/Documents/speed-restriction-air-quality-report-2013-for-web.pdf
28 Directive 2008/50/EC on ambient air quality and cleaner air for Europe
29 United States Environmental Protection Agency, "Particulate Matter (PM) Basics", https://www.epa.gov/pm-pollution/particulate-matter-pm-basics
30 Air Quality Expert Group (2007), "Trends in Primary Nitrogen Dioxide in the UK," https://uk-air.defra.gov.uk/assets/documents/reports/aqeg/primary-no-trends.pdf
31 European Commission Press release (17 May 2018), "Air quality: Commission takes action to protect citizens from air pollution," http://europa.eu/rapid/press-release_IP-18-3450_en.htm
32 European Commission Press release (15 February 2017), "Commission warns Germany, France, Spain, Italy and the United Kingdom of continued air pollution breaches," http://europa.eu/rapid/press-release_IP-17-238_en.htm
33 World Health Organization, Global Health Observatory data repository, http://apps.who.int/gho/data/node.main.152?lang=en
34 Data extracted from Department for Environment, Food and Rural Affairs, "UK Air: Air Information Resource," https://uk-air.defra.gov.uk/
35 Krotkov, McLinden, Li, Lamsal, Celarier, Marchenko, Swartz, Bucsela, Joiner, Duncan, Boersma, Veefkind, Levelt, Fioletov, Dickerson, He, Lu and Streets, "Aura OMI observations of regional SO2 and NO2 pollution changes from 2005 to 2015," Atmos. Chem. Phys., 16, 4605–4629, 2016. http://www.atmos-chem-phys.net/16/4605/2016/
36 Data from European Commission, "Emissions Database for Global Atmospheric Research," http://edgar.jrc.ec.europa.eu/
37 NASA Ozone Watch, "What is a Dobson Unit?," https://ozonewatch.gsfc.nasa.gov/facts/dobson.html
38 The Independent, 21 May 2013, "Joe Farman: Scientist who first uncovered the hole in the ozone layer," http://www.independent.co.uk/news/obituaries/joe-farman-scientist-who-first-uncovered-the-hole-in-the-ozone-layer-8624438.html

39 The Guardian, 18 April 2015, "Thirty years on, scientist who discovered ozone layer hole warns: 'it will still take years to heal',". https://www.theguardian.com/environment/2015/apr/18/scientist-who-discovered-hole-in-ozone-layer-warns
40 Farman, J.C.; Gardiner, B.G.; Shanklin, J.D., "Large losses of total ozone in Antarctica reveal seasonal ClOx/NO_x interaction," Nature 315, 207–210 16 May 1985
41 NASA, 10 December 2001, "Research Satellites For Atmospheric Sciences,1978-Present,"https://earthobservatory.nasa.gov/Features/RemoteSensingAtmosphere/remote_sensing5.php
42 Photograph "View of part of the Los Angeles Civic Center masked by smog in 1948," Los Angeles Times Photographic Archives (Collection 1429). Library Special Collections, Charles E. Young Research Library, UCLA
43 Pitts, J.N. and Stephens, E.R., "Arie-Jan Haagen-Smit 1900 – 1977," Journal of the Air Pollution Control Association, 1978, pp. 516–517
44 Proctor R.N., "The history of the discovery of the cigarette–lung cancer link: evidentiary traditions, corporate denial, global toll," Tobacco Control 2012;21:87–91
45 The Independent, 10 May 2008, "The complete guide to: the Isle of Man," https://www.independent.co.uk/travel/uk/the-complete-guide-to-the-isle-of-man-824988.html
46 The Times, 19 June 2016, "Nautical stripes and shipshape accessories equal summery seaside chic," https://www.thetimes.co.uk/article/the-cold-sea-rns8fk7m0
47 Air Quality Expert Group, 2009, "Ozone in the United Kingdom," https://uk-air.defra.gov.uk/library/aqeg/publications
48 Source: http://www.ems.psu.edu/~brune/m532/m532_ch4_troposphere.htm
49 State of Global Air 2017, "A special report on global exposure to air pollution and its disease burden," https://www.stateofglobalair.org/report
50 Norwegian Environment Agency, 9 April 2018, "Acid Rain," http://www.environment.no/Topics/Air-pollution/Acid-rain/
51 European Environment Agency, 16 October 2018, "Emissions of the main air pollutants in Europe," https://www.eea.europa.eu/data-and-maps/indicators/main-anthropogenic-air-pollutant-emissions/assessment-4
52 Photograph: the author
53 Photograph: http://www.geograph.org.uk/photo/5092802 licensed for reuse under a Creative Commons Licence.
54 Natural England, "Views About Management: A statement of English Nature's views about the management of Goss and Tregoss

Moors Site of Special Scientific Interest (SSSI)," https://necmsi.esdm.co.uk/PDFsForWeb/VAM/1001443.pdf

55 Eurostat, "Archive: Agricultural census in the United Kingdom," https://ec.europa.eu/eurostat/statistics-explained/index.php?title=Archive:Agricultural_census_in_the_United_Kingdom

56 Defra (2002), "Ammonia in the UK," http://adlib.everysite.co.uk/resources/000/109/544/ammonia_uk.pdf

57 Images from Wikimedia https://commons.wikimedia.org/wiki/File:%22Fumifugium%22,_Evelyn_Wellcome_L0009664.jpg and Willem van de Poll, https://commons.wikimedia.org/w/index.php?curid=66648059 Copyrighted work available under Creative Commons Attribution only licence CC BY 4.0 http://creativecommons.org/licenses/by/4.0/

58 Janner, M., "The politics of London air: John Evelyn's Fumifugium and the Restoration," The Historical Journal, 38, 3, pp. 535–551, 1995

59 "The Big Smoke: Fifty years after the 1952 London Smog," 2005, edited by: Virginia Berridge and Suzanne Taylor, © Centre for History in Public Health, London School of Hygiene & Tropical Medicine

60 Wilkins, E.T., "Air pollution aspects of the London fog of December 1952," Q J R Meteorol Soc. 1954;80:267–71

61 Bell, M.L.; Davis, D.L. & Fletcher, T. (2004), "A Retrospective Assessment of Mortality from the London Smog Episode of 1952: The Role of Influenza and Pollution," Environ Health Perspect. 112 (1 January): 6–8

62 Adapted from Richard P. Turco, "Earth Under Siege: From Air Pollution to Global Change," Oxford University Press, 1997

63 News stories from BBC website: https://www.bbc.co.uk/news/health-35629034; https://www.bbc.co.uk/news/world-europe-38078488; https://www.bbc.co.uk/news/science-environment-35568249; https://www.bbc.co.uk/news/health-37483616

64 COMEAP, "The Mortality Effects of Long-Term Exposure to Particulate Air Pollution in the United Kingdom," 2010 and "Associations of long-term average concentrations of nitrogen dioxide with mortality," 2018

65 Royal College of Physicians and Royal College of Paediatrics and Child Health, "Every breath we take: the lifelong impact of air pollution," February 2016

66 Pharmacy Magazine, 2017, "Trouble ahead as hayfever incidence rockets," https://www.pharmacymagazine.co.uk/trouble-ahead-as-hay-fever-incidence-rockets

67 Allergy UK, "Statistics," https://www.allergyuk.org/information-and-advice/statistics, accessed January 2019

68 Photograph source: http://www.ebay.ca/itm/HORIBA-Analyzer-Unit-NO_x-SO2-O2-/112383251759?hash=item1a2a90552f:g:N8AAA-OSw65FXuP9n
69 Department for Environment, Transport and the Regions (undated), "UK Air Pollution: Winter Smog Episodes," https://uk-air.defra.gov.uk/assets/documents/reports/empire/brochure/winter.html
70 Defra (May 2017), "Draft UK Air Quality Plan for tackling nitrogen dioxide: Technical Report," https://consult.defra.gov.uk/airquality/air-quality-plan-for-tackling-nitrogen-dioxide/supporting_documents/Technical%20Report%20%20Amended%209%20May%202017.pdf
71 King's College London, November 2018, "London Air Quality Network: Summary report 2017"
72 The Nation, 2 March 2000, "The Secret History of Lead," https://www.thenation.com/article/secret-history-lead/
73 United Nations Environment Programme, "Leaded Petrol Phase Out: Global Status as at June 2016," http://staging.unep.org/Transport/new/PCFV/pdf/Maps_Matrices/world/lead/MapWorldLead_June2016.pdf
74 Mother Nature Network, 5 October 2015, "Here's how VW's diesel 'defeat devices' worked," https://www.mnn.com/green-tech/transportation/blogs/heres-how-vws-diesel-defeat-devices-worked
75 Oldenkamp, van Zelm, Huijbregts, "Valuing the human health damage caused by the fraud of Volkswagen," Environmental Pollution Volume 212, May 2016, Pages 121–127, http://www.sciencedirect.com/science/article/pii/S0269749116300537
76 Information taken from Defra, "National Atmospheric Emissions Inventory," http://naei.defra.gov.uk/
77 World Health Organization Regional Office for Africa, "Air pollution," https://afro.who.int/health-topics/air-pollution
78 The Sun, https://www.thesun.co.uk/news/2655590/london-smog-toxic-air-pollution-heatwave-health-sadiq-khan/
79 Image used by permission of Clean Air Now, Claire Matthews, http://claire-matthews.com/, and Vasilisa Forbes, https://vasilisaforbes.co.uk
80 "In Praise of Air" used by permission of the author. Non-exclusive world rights in perpetuity, only within the volume as described, not to be reproduced separately without further agreement, full author credit, copyright to remain with the author.
81 How Many Left, statistics about every make and model of vehicle registered in the UK, https://www.howmanyleft.co.uk/make/volkswagen accessed January 2019

82 UK Air Data Archive, available at https://uk-air.defra.gov.uk/data/
83 Environment Analyst (2015), "Ricardo-AEA to quality control UK air monitoring network" https://environment-analyst.com/30161/ricardo-aea-to-quality-control-uk-air-monitoring-network
84 Figures derived from data provided by Windfinder.com, used by permission. Accessed from: https://www.windfinder.com/windstatistics/birmingham and https://www.windfinder.com/windstatistics/edinburgh
85 Image taken from Google Maps
86 Information derived from Land Registry data taken from the rightmove website http://www.rightmove.co.uk/
87 The Independent, 13 October 2015, "Can a street's smell significantly reduce its house prices?," http://www.independent.co.uk/property/house-and-home/property/can-a-streets-smell-significantly-reduce-its-house-prices-a6692831.html
88 Ben Johnson for Historic UK, "The Great Horse Manure Crisis of 1894," http://www.historic-uk.com/HistoryUK/HistoryofBritain/Great-Horse-Manure-Crisis-of-1894/
89 Dravnieks, A.; Masurat, T.; Lamm, R.A., "Hedonics of Odours and Odour Descriptors": in Journal of the Air Pollution Control Association, July 1984, Vol. 34 No. 7, pp. 752–755
90 Chartered Institute of Environmental Health, "Analysis of Defra's statutory nuisance survey results by CIEH," December 2011. Available from http://randd.defra.gov.uk/Default.aspx?Menu=Menu&Module=More&Location=None&ProjectID=16101&FromSearch=Y&Publisher=1&SearchText=cieh&SortString=ProjectCode&SortOrder=Asc&Paging=10#Description
91 Photograph used by permission of Breckland Industrial Ltd (http://www.brecklandindustrial.co.uk/Cleaning-Services/Ductwork-Cleaning/)
92 Photographs by permission of Agriculture & Horticulture Development Board (www.ahdb.org.uk)
93 Photograph: the author
94 Photograph used by permission of i2 Analytical UK Ltd, https://www.i2analytical.com/services/i2-hanby-gauge/
95 Hodge, Jones and Allen Solicitors, 11 January 2019, "Attorney-General moves to quash inquest of nine-year-old girl," https://www.hja.net/press-releases/attorney-general-moves-to-quash-inquest-of-nine-year-old-girl/
96 BBC "So I Can Breathe," https://www.bbc.co.uk/news/science-environment-38853910
97 Depa, P.M.; Dimri, U.; Sharma, M.C.; Tiwari, R., "Update on epidemiology and control of Foot and Mouth Disease – A menace to

international trade and global animal enterprise," Vet. World, 2012, Vol.5(11): 694–704

98 Ghosh R.E., Freni-Sterrantino A., Douglas P., Parkes B., Fecht D., de Hoogh K., Fuller G., Gulliver J., Font A., Smith R.B., Blangiardo M., Elliott P., Toledano M.B., Hansell A.L., "Fetal growth, stillbirth, infant mortality and other birth outcomes near UK municipal waste incinerators; retrospective population based cohort and case-control study," Environment International, Volume 122, January 2019, Pages 151–158

99 Tang, T.; Fujita, T.; Tanhata, T.; Minowa, M.; Doi, Y.; Kato, N.; Kunikane, S.; Uchiyama, I.; Tanaka, M.; Uehata, T., "Risk of adverse reproductive outcomes associated with proximity to municipal solid waste incinerators with high dioxin emission levels in Japan," J Epidemiol. 2004 May;14(3):83-93.

100 Friends of the Earth, http://www.foeeurope.org/incineration, accessed August 2017

101 UK Without Incineration Network, quoted by Global Alliance for Incinerator Alternatives, http://www.no-burn.org/europe-members/

102 Based on data from National Atmospheric Emissions Inventory, http://naei.defra.gov.uk/

103 Letsrecycle.com, 11 September 2017, "Rivenhall EfW granted Environmental Permit", https://www.letsrecycle.com/news/latest-news/rivenhall-efw-granted-environmental-permit/

104 Box, G. E. P. (1979), "Robustness in the strategy of scientific model building," in Launer, R. L.; Wilkinson, G. N., Robustness in Statistics, Academic Press, pp. 201–236

105 Lavoisier, A. translated by Robert Kerr (1790), "Elements of chemistry in a new systematic order, containing all the modern discoveries," https://books.google.co.uk/books?id=4B8UAAAAQAAJ&dq=editions:__o3stkbE5EC&hl=de&pg=PR3&redir_esc=y#v=onepage&q&f=true

106 Adams, S. (2014), "Coal Rolling is the New Old Black," https://energypast.com/2014/08/11/coal-rolling-is-the-new-old-black/ quoting The Atlantic Monthly, August 1867, "Cincinnati," James Parton

107 Photograph: The author

108 Adapted by the author from: Briggs, G. A. (1965), A plume rise model compared with observations. JAPCA, 15, 433–438. Briggs, G. A. (1968), CONCAWE meeting: Discussion of the comparative consequences of different plume rise formulas. Atmospheric Environment, 2, 228–232

109 Photographs: (a) Philip Lambert http://www.choppedonion.com, used by permission; (b) The author

110 Bosanquet, C. H., and Pearson, J. L., "The Spread of Smoke and Gases from Chimneys," Trans. Faraday Soc., 32, 1249 (1936)
111 Sutton, O. G., "The Theoretical Distribution of Airborne Pollution from Factory Chimneys," Quart. J. Roy. Meteorol. Soc., 73, 426 (1947)
112 Photograph: the author
113 Data taken from UK Air Information Resource, https://uk-air.defra.gov.uk
114 Population of Eskdalemuir in 2001 taken from Wikipedia, https://en.wikipedia.org/wiki/Eskdalemuir
115 World Health Organization Regional Office for Europe, Copenhagen (2000), "Air Quality Guidelines for Europe," Second Edition
116 Figure prepared by Jo Green based on data taken from UK Air Information Resource, https://uk-air.defra.gov.uk using tools available at The OpenAir Project, http://www.openair-project.org/
117 Adapted from Netcen for UK Government (2006), "UK Smoke and Sulphur Dioxide Network 2004"
118 NASA "Air Quality Observations From Space," https://airquality.gsfc.nasa.gov/
119 IPCC, 2013: Summary for Policymakers. In: "Climate Change 2013: The Physical Science Basis. Contribution of Working Group I to the Fifth Assessment Report of the Intergovernmental Panel on Climate Change," Stocker T.F., Qin D., Plattner G.-K., Tignor M., Allen S.K., Boschung J., Nauels A., Xia Y., Bex V. and Midgley P.M. (eds.), Cambridge University Press, Cambridge, United Kingdom and New York, NY, USA.
120 Defra, "UK Plan for tackling roadside nitrogen dioxide concentrations," Technical Report, July 2017
121 Oxfordshire Air Quality, Local Air Quality Management – South Oxfordshire, https://oxfordshire.air-quality.info/local-air-quality-management/south-oxfordshire
122 Public Health England, Public Health Outcomes Framework, https://data.england.nhs.uk/dataset/phe-indicator-30101
123 AP Images used by permission, http://www.apimages.com/metadata/Index/China-Pollution/38403c6188824338adf197311339c5a7/41/0
124 Xinhuanet, 20 February 2017, "China criticizes several cities' response to air pollution," http://news.xinhuanet.com/english/2017-02/20/c_136068637.htm
125 Health Effects Institute (2018), "State of Global Air," https://www.stateofglobalair.org/report
126 Digital India, "Road Transport Year Book: 2013–14 and 2014–15," available from https://data.gov.in/catalog/road-transport-year-book-2013-14-and-2014-15

127 Hindustan Times, 1 November 2016, "Delhi records worst air quality in three years this Diwali," http://www.hindustantimes.com/delhi-news/delhi-records-worst-air-quality-in-three-years-this-diwali/story-OuqsDMSUiKT9HxOlm9yTyN.html
128 Care2, 5 July 2017, "10 Cities With the Cleanest Air in the World," http://www.care2.com/causes/10-cities-with-the-cleanest-air-in-the-world.html
129 Information taken from records compiled by Weatherbase, http://www.weatherbase.com/

ACKNOWLEDGEMENTS

This book originated more than twenty years ago, when I was working for the late Martin Tasker at ICI Technology. Martin suggested that he and I should co-author a "Dispersion modelling guide," which eventually evolved into this book. It took me a while, but I'm very grateful to Martin for sowing the seed of the idea and showing me how to deal with industrial process emissions. Before and since then, I have benefited from the expertise and understanding of too many colleagues to mention individually, who have had to put up with my slightly obsessive interest in air quality, and fill in the gaps in my understanding of every other area of environmental science. My thanks to every one of these long-suffering colleagues.

Talking of long-suffering, I took a four-month sabbatical from Ricardo Energy and Environment to write this book. My thanks to Sean Christiansen and Beth Conlan, and to everyone who concealed any resentment they might reasonably have felt at my absence to do my job on top of their own, and demonstrate that I'm a bit less indispensable than I might have thought. I'm grateful to Professor Duncan Laxen for taking the time to provide thoughtful and constructive feedback. Any shortcomings in the book are, of course, all my own responsibility.

Thank you to my agent Joanna Swainson and publishers Abbie Headon and Matt Casbourne, for steering me through the arcane and unfamiliar world of publishing, and ensuring that the worst of the dad-jokes never saw the light of day.

And finally, thank you to my sons Matt, Jonny and Ben, who have had to put up with obscure references to air pollution and occasional diversions to look at chimneys over many years. And to my wife Emma, who (quite apart from going out to work to pay the bills while I was speculatively writing this) has been constantly encouraging and supportive, continued to laugh at my jokes, and still gives the impression that air quality is almost as fascinating to her as it is to me, and a great subject for a book.

INDEX

Acid deposition 89, 91
Acid rain 5–7, 57, 88, 89, 91, 93, 95, 97, 99, 292
Agatha Christie 97
Agricultural waste 137–9
Agriculture 37, 44, 53, 96, 98, 135, 138, 175, 182, 221, 295
Air monitoring station 143, 223
Air pollutant 33, 37–41, 43, 44, 46, 57, 58, 89, 91, 92, 109, 110, 112, 113, 119, 129, 134, 141, 150, 156, 188, 190, 191, 195, 210, 216, 224, 236, 239, 241, 252, 253, 255, 266, 272–4, 292
Air pollution 2–4, 7–9, 29, 34–6, 38–46, 54, 57, 76, 88–9, 91–5, 97–101, 103–4, 109–14, 118–20, 125, 131–34, 138–42, 144–51, 157, 159, 162, 165, 167, 180–81, 185–95, 203, 206–7, 212, 216–17, 221, 223–5, 232–3, 235, 237–8, 241, 244, 249, 251–2, 254–5, 263–77, 279–88
Air pressure 13, 21, 27
Air quality
Air Quality Expert Group 53, 291, 292
Air quality guideline 297
Air Quality Management Area (AQMA) 272
Air quality standard 36, 52, 54–7, 62, 111, 118–22, 130, 143, 146–7, 210, 212, 224, 239–40, 242, 244, 246–7, 258, 267, 269, 273–4
Aircraft 27, 133, 134, 232
Airpocalypse 142, 143, 279
Airshed 48, 49, 78
Alchemy 18, 101, 218
Aleix Segura Vendrell 188
Alex Rhys-Taylor 164
Alkali Act 266
Allergic rhinitis 113
Ammonia (NH3) 97
Anaerobic digestion 63, 139
Antoine Lavoisier 154, 218

Argon (Ar) 14, 17, 18, 29, 30, 33, 63
Arie Jan Haagen-Smit 75, 78
Aristotle 217
Arthur Conan Doyle 105
Ash 63, 207
Asian Development Bank 190, 280
Asthma 81, 113, 187, 189, 199, 247
Atmosphere 1–2, 4–37, 40, 46, 48–9, 51–5, 57, 59–60, 63, 65–71, 73–4, 76–7, 79–80, 82–3, 85–6, 88, 91–2, 95–9, 101, 103, 105, 123, 127–8, 133–4, 143, 145, 155–6, 159–60, 167–9, 173, 181, 184, 186–8, 196, 201, 204–7, 217, 219–22, 225, 227, 233, 237, 240–41, 255, 258, 265–7, 270–1, 288
Atmospheric chemistry 30, 53, 81, 217, 218, 220
Austin Bradford Hill 108
BBC 25, 109, 110, 118, 190, 279, 293, 295
Beaufort scale 232, 252
Beijing 62, 191, 279, 280
Benzene 123, 124, 173
Beta Attenuation Monitor (BAM) 261, 262
Bill Bryson 123
Biodiversity 90, 92, 95
Biogas 63, 139
Biomass 40, 63, 271, 282
Bloomsbury 246, 247, 250, 251–3
Body Shop 142, 143, 145, 279
Bonfire 208, 282
Brian Gardiner 71
British Antarctic Survey 69, 71–3
C.H. Bosanquet 240, 297
California 56, 75, 78, 79
California Air Resources Board 78
Cambridge 4, 5, 11, 66, 90, 115, 163, 164, 297
Carbon (without monoxide/dioxide)
Carbon dioxide (CO2)
Carbon monoxide (CO) 41–3, 57, 77, 117, 139, 140, 258
Carboniferous 19
Cardiovascular disease 112, 276
Carl Friedrich Gauss 226, 241
Carl Scheele 220
Carmaggedon 142
Catalyst 41, 43, 53, 123, 244, 257
Catalytic convertor 41–4, 53, 57, 79, 123, 124, 139, 274
CFCs 70–2, 74, 140
Chai Jing 191
Charles Dickens 105
Charles II 103
Chartered Institute of Environmental Health (CIEH) 170, 295
Chemistry 4–6, 18, 30, 32, 53, 54, 64, 66, 68, 75, 81, 101, 172, 173, 217–21, 296
Chimney 7, 94, 132, 156, 159, 196, 199, 210, 225, 226, 233, 235–7, 241, 254
China 56, 60, 61, 135, 137, 138, 142, 190, 191, 278–82, 297
Chlorofluorocarbons 26, 70
Christian Schönbein 80

Chronic obstructive pulmonary disease 112
Claude Monet 105
Clean Air Act 108, 266
Clean Air in London 149
Clean Air Strategy 132
Climate change 3, 8, 29, 33–7, 39, 40, 44, 46, 54, 73, 74, 86, 149, 194, 217, 271, 290, 297
Coal 19, 33, 59-63, 88, 101, 103, 105-106, 109, 128, 138, 140, 187, 207, 221-222, 224, 241, 259-260, 266, 278-279, 282, 286
Combined Heat and Power (CHP) 103, 237
Committee on the Medical Effects of Air Pollutants (COMEAP) 110
Community Multi-scale Air Quality (model) 241
Congestion charge 45, 121
Convention on Long-range Transboundary Air Pollution (CLRTAP) 36, 97
Covalent 30, 66
Crops 39, 63, 68, 75–9, 90, 98, 138, 206
Daniel Rutherford 220
Defeat device 129
Department for Environment, Food and Rural Affairs (DEFRA) 198, 291, 294
Department of Transport 114, 115
Dephlogisticated air 18, 220
Diesel 43, 53, 103, 104, 122, 128, 129, 131, 135, 140, 142, 193, 194, 279, 286
Diffusion tube 242
Digestate 139
Dimethyl sulphide 172
Dioxins / Dioxins and furans 201–10, 259
Dispersion 36, 156, 160, 178, 225–37, 239–41, 267, 284, 299
domestic wood combustion 131
Drosera 90
Dust 2, 12, 14, 27-28, 44, 49, 103, 180–6, 221, 262, 283
Early death 3, 4
Edward Morley 218
Electric vehicle 85, 104, 194–6, 274, 288
Electronic nose 176
Ella Kissi-Debrah 189
Emissions Ceiling Directive 97
Environment Act 119
Environment Agency 93, 103, 119, 135, 150, 210, 266, 292
Environmental Permitting 133
Epidemiology 107, 200, 295
Eric Cantona 117
Eskdalemuir 222, 246, 247, 251, 252, 297
European Commission 55, 205, 243, 259, 273, 291
European Court of Justice 55
European Environment Agency 135, 150, 292
Eutrophication 91, 96
Exoplanet 1, 11, 289
Frank Pasquill 240
Friends of the Earth 203, 296
Frisbee gauge 183
Fumifugium 101, 102, 187, 293

Gaussian 226–8, 230–3, 240, 241
Geothermal 12, 14, 63
GJ 1132 1, 2, 8, 11–3
Global warming 2, 54, 123, 134, 290
Graham Sutton 240
Great Stink 165–7
Greenhouse gas 15, 16, 33, 35, 36, 37, 39, 40, 43–5, 133, 134, 274, 290
Greenpeace 203, 206
Ground source heat pumps 63
Habitat site 92, 95
Hay fever 113, 114, 293
Health Effects Institute 86, 281, 297
Healthy Air 149
Heart disease 112, 138
Hedonic tone 168, 171
Helium (He) 13, 30–2
Henry Cavendish 220
Horse manure 165, 295
Hybrid car 194
Hydro 39, 63
Hydrogen (H2) 12, 13, 19, 20, 30, 31, 42, 64, 65, 76, 77, 91, 97, 172, 174, 187, 195, 196, 219, 255, 257
Hydrogen sulphide (H2S) 12, 13, 172, 174, 187, 257
Hydroxyl (OH) 76, 77
Ice 13, 15, 27
Incineration 151, 196–209, 214, 216, 259, 296
Independent 81, 125, 164, 198, 210, 291, 292, 295
India 56, 60, 61, 87, 137, 193, 278, 281–3, 297
Industrial Emissions Directive 209
Infra-red 25
Intergovernmental Panel on Climate Change (IPCC) 271, 297
Ionosphere 2, 23, 28
Isaac Asimov 220
J.L. Pearson 240, 297
James Hansen 35
James Michelson 218
James Parton 221, 296
Jan Baptist van Helmont 18, 32
Jeremy Clarkson 195
Joanna Gavins 144
Joe Farman 71, 291
John Evelyn 101–5, 187, 221, 266, 293
John Gummer 119
Jonathan Shanklin 71, 74
Joseph Bazalgette 166
Joseph Priestley 218, 219
Jupiter 12–4
Landfill 169, 177, 199, 241
Laughing gas 54
Lead 122–5, 139, 140, 149, 266, 294
Lichen 92, 95, 96
Lindvall Hood 177, 178
LondonAir 149, 150
Low emission zone 121, 122, 239
Low speed zone 46
Lung 49, 112, 138, 140, 146, 189, 200, 292
Lung cancer 112, 138, 189, 200, 292
Mario Molina 70
Mars 4, 14, 15, 189, 191
Marsh Fritillary butterfly 94

Marylebone Road 222, 223, 248, 249, 250, 251
Max Planck Institute 12, 289
Mercury (planet) 15, 219
Mesosphere 2, 23, 27, 28
Meteorology 218
Meteors 27, 28
Methane (CH4) 12, 13, 35
Methyl mercaptan 172
Micro-organism 6
Montreal Protocol 73
Mortality 106, 110, 111, 112, 125, 187, 189, 200, 276, 277, 290, 293, 296
Moss 91, 92, 111
NASA 6, 35, 71–3, 264, 290–2, 297
National Air Quality Strategy 119
National Health Service 131
Natural gas 31, 62, 63, 109, 254
Nepal 21, 22, 56, 287
New York 159, 165, 289, 297
Nitric oxide (NO) 83, 244
Nitrogen (N2)
Nitrogen deposition 89–91
Nitrogen dioxide (NO2) 244
Nitrous oxide (N2O) 35, 53, 54
Northern Lights 28
Norway 88, 89
Norwegian Environment Ministry 88
Nos 54
NOx box 244, 262
Nutrient nitrogen 90, 94
Obesity 4, 29, 147, 189, 191, 199, 290
Odour 5, 75, 80, 103, 113, 164, 165, 167–79 , 185, 186, 257, 258, 295
Odour Unit 175
Oil 19, 59, 62, 109, 127, 128, 134, 136, 206, 224, 259
Olfactometry 174–6
OpenAir 252, 254, 297
Oxides of nitrogen (NOx) 41, 44, 51–5, 74, 78, 82, 83, 89, 91, 95, 135, 140, 210, 222, 223, 238, 244, 247–50, 252–55, 286
Oxides of sulphur (SOx)
Ozone (O3) 259, 270, 279, 291, 292
Ozone hole 5, 71, 72, 74, 81, 125
Particulate matter 44, 46, 49, 50, 51, 74, 87, 100, 110, 125, 127, 128, 131, 132, 135, 136, 140, 143, 183, 184, 194, 202, 221, 222, 259, 260–3, 274, 278, 283, 291
Passive smoking 4, 29, 147, 189, 191
Paul Crutzen 70
Phlogiston 18, 219, 220
Photochemical ozone 5
Photochemistry 64
Photon 68, 69
Photosynthesis 19, 20, 64, 181, 219, 270
$PM_{0.1}$ 50
PM_1 49, 260, 263
PM_{10} 49, 50, 111, 112, 126, 127, 134, 136, 137, 254, 260–3, 282
$PM_{2.5}$ 49–51, 55–6, 110–2, 125–8, 130–8, 189, 200, 222, 239, 260–3, 275–8, 281, 282
Power station 20, 51, 59, 60, 160, 195
Prevailing wind 156, 157, 159, 162–5

Public Health Outcomes Framework 276, 277, 297
Reduced sulphur 5
Refrigerants 70, 71
Rendering 176
Richard Doll 108
Road traffic 41–4, 83, 89, 92, 100, 104, 121, 156, 195, 232, 237–9, 248, 268, 282, 284,
Robert Boyle 101, 218
Royal College of Paediatrics and Child Health 130, 293
Royal College of Physicians 130, 293
Russ Abbbot 48
Sadiq Khan 122, 142, 294
Satellite 71, 72, 263–5
Saturn 12, 13
Sean Adams 221
Sewage 113, 165–7, 169–171, 174, 177, 179, 185, 257
Sherwood Rowland 70
Shipping 58, 127, 133–6
Shooting stars 23, 27
Simon Armitage 145
Smell 8, 13, 29, 75, 76, 79, 80, 81, 97, 101, 113, 164–9, 171–7, 179, 180, 186, 187, 235, 257, 270, 295
Smog 74–6, 78, 81, 105–8, 140, 145, 191, 266, 283, 292–4
Smoke 103, 105, 106, 108, 109, 132, 138, 139, 140, 167, 176, 181, 184, 187, 221, 222, 232, 240, 255, 256, 259, 260, 263, 279, 293, 297
Snot 181
Solar 1, 2, 11, 12, 15, 28, 39, 63, 67, 288, 290
Southern Lights 23, 28
Stable (atmospheric conditions) 28, 34, 54, 65–7, 72, 228, 229, 240
Stack 94, 160, 210, 225, 228, 233, 236, 237
Standard deviation 227, 228, 230, 233, 235
Sting 9, 289
Stratosphere 2, 5, 23, 25–7, 66, 68–70, 72, 73, 81, 82, 86
Stratospheric ozone 5, 26, 32, 68–70, 72–4, 82
Straw 127, 136
Sulphur dioxide (SO2) 38, 57–63, 88, 89, 91, 106, 108, 109, 127, 134, 139, 140, 211, 223, 224, 254, 255–7, 260, 267, 279, 281, 297
Sunday Times 81
Sundew 91
Sunlight 22, 25, 26, 29, 51, 52, 64, 65, 67–9, 76–8, 81–3, 87, 105, 113, 248
Surface roughness 20
Sweden 284
Sydney Chapman 66
Tapered Element Oscillating Microbalance (TEOM) 206
Tax 45, 194, 218, 220
Temperature 12, 14, 15, 23–8, 33, 36–8, 68, 159, 201, 237, 285
The Institute for Health Metrics and Evaluation 86
The Police 9, 48
The police 116–7
The Sun 14, 15, 22, 23, 28, 65, 67, 69, 81, 82, 91, 140–2, 155, 249, 279, 294

Theresa May 141
Thermosphere 2
Thomas Midgley 122
Tony Ryan 144
Tropopause 24, 26, 68
Troposphere 2, 23–6, 68, 74, 82, 159, 160, 265, 292
Twitter 106, 146, 148
UK Without Incineration Network (UKWIN) 203, 296
Ultra Low Emission Zone 122
Ultra-violet 69
Uncertainty 112, 192, 193, 214, 215, 234, 236, 274
United Nations 36, 97, 124, 290, 294
United Nations Environment Programme (UNEP) 124, 294
United States 31, 56, 85, 90, 137, 286, 291
United States Environmental Protection Agency (EPA) 85, 291
University of Sheffield 144, 145
Unleaded petrol 124
Unstable (atmospheric conditions) 228, 229, 240
Venus 2, 8, 12, 15, 16, 90, 289
Venus Flytrap 90
Volatile organic compounds 41, 43, 74, 84, 140, 258
Volkswagen 53, 128, 129, 147, 294
Volvo 194, 274
Waste to energy 93, 94, 151, 152, 196–9, 201–3, 206, 209–211, 213–5
Water 4, 12, 13, 15, 16, 20, 22, 29, 32, 33, 37, 42, 48, 63, 64, 76, 77, 83, 85, 89, 94, 97, 114, 116, 117, 129, 147, 148, 166, 181, 184, 189, 191, 199, 206, 217, 219, 241, 257, 259, 261, 266, 271, 290
Water pollution 4, 29, 94, 147, 191
Water vapour 12, 16, 63, 259
William Ramsay 29
Wood 33, 39, 40, 63, 96, 127, 128, 131–3, 136, 138, 207, 219, 271, 275, 278, 282
World Health Organization (WHO) 29, 124, 138, 148, 281, 283, 290, 291, 294, 297
York 6, 159, 168, 173, 289, 297